One-Page Summaries for Algebra, Geometry, and Pre-Calculus

By Kathryn Paulk

Copyright © 2021

Updated: 02/02/2024

TABLE OF CONTENTS

INTRODUCTION

This book contains a set of one-page summaries and problem examples of important math topics covered in algebra, geometry, and pre-calculus. Current students may use the summaries to reinforce the material learned in class. The concentrated, one-page math summaries give students an excellent overview of the material and may provide new insights. Students planning to take SAT exams or higher-level math courses may use the one-page summaries for review and reference.

Part I contains a set of one-page math summaries.
The list of one-page summaries is included on the next page.

Part 2 contains example math problems and solutions for the summaries. Examples for polynomial division (both long division and synthetic division) are also included in Part 2.

#	One-Page Summary	Contents
1.	Lines	Perimeter, Circumference
2.	Area	Area of Rectangles, Circles, Ovals
3.	Volume	Vol of Cylinders, Cones, Pyramids
4.	Regular Polygons	Area of Regular Polygons
5.	Two Points	Midpoint, Dist., Slope, Linear Eqn.
6.	Linear Equations	Slope Intercept, Point Slope, Std.
7.	Quadratic Eqns.	Standard, Vertex, Factored Forms
8.	Circles	Lengths of chords and tangent lines
9.	Unit Circle	Sine and Cosine
10.	Triangles	Law of Sines and Law of Cosines
11.	Exponents	Exponential Properties
12.	Factoring	Squares, Cubes, Binomial Eqn.
13.	Exponential Eqns.	Exponential Growth and Decay
14.	Logs	Logarithmic Properties
15.	Parent Functions	12 Parent Functions
16.	Translations	Vertical and Horizontal Translations
17.	Graph Polynomials	Zeros, Multiplicity, End Behavior
18.	Graph Rational Fun.	Vertical, Horiz. & Slant Asymptotes
19.	Conic Sections	Parabola, Ellipse, and Hyperbola
20.	Series	Arithmetic and Geometric Series
21.	Basic Stats	Mean, Median, Mode, Std. Dev.
22.	Vectors	Dot Product, Cross Product
23.	Complex Numbers	$a + bi$, Rectangular to Polar Form
24.	Polar Curves	Spiral, Rose, Limacon

PART 1 – SUMMARIES

The following sections contain one-page summaries of important math topics covered in algebra, geometry, trigonometry, pre-calculus, and college algebra. Problem examples are included in Part 2.

These summaries are for students who are familiar with the topics. They should be used for review and reference.

Lines

Perimeter	a □ b	$P = 2a + 2b$
Circumference	r (circle)	$C = 2\pi r$ $\pi \approx 3.14$
Pythagorean Theorem	c b a (right triangle)	$a^2 + b^2 = c^2$
$30 - 60 - 90°$ Triangle	h 60 30 90 $\left(\frac{\sqrt{3}}{2}\right)h$	$\left(\frac{1}{2}\right)h$
$45 - 45 - 90°$ Triangle Isosceles Right Triangle	h 45 45 90 $\left(\frac{\sqrt{2}}{2}\right)h$	$\left(\frac{\sqrt{2}}{2}\right)h$

Area

Rectangle & Parallelogram $A = bh$	$b \qquad b \qquad b \qquad h$
Triangle $A = \left(\frac{1}{2}\right)bh$	$b \qquad b \qquad b \qquad h$
Trapezoid $A = \left(\frac{b_1 + b_2}{2}\right)h$	$b_2 \qquad b_2 \qquad b_2$ $b_1 \qquad b_1 \qquad b_1 \qquad h$
Circle $A = \pi \cdot r^2$	$r \qquad \pi \approx 3.14$
Oval $A = \pi \cdot r_1 r_2$	$r_2 \qquad r_1$

Volume

Equations	Circular Base	Rectangular Base	Triangular Base
Top = Base and is Parallel $V = BH$	Cylinder		
Top = Point $V = \left(\frac{1}{3}\right) BH$	Cone	Pyramid	Pyramid
Sphere $V = \left(\frac{4}{3}\right) \pi r^3$		$r = radius$ $\pi \approx 3.14$	
Notes	B = Base Area H = Height (How tall?)		

Regular Polygons

Definition	Regular Polygons have "n" equal sides.
Measurement of each Interior Angle	$\theta = \dfrac{360°}{n}$
Perimeter $n = \#\ sides$ $s = side\ length$	$P = n \cdot s$
Radius	$r =$ Distance from Center to Corner
Apothem	$a =$ Distance from Center to Flat Edge
Area Of Triangle	$A = \left(\dfrac{1}{2}\right)(base)(height)$ $A = \left(\dfrac{1}{2}\right)\left(\dfrac{s}{2}\right)a$ $A = \left(\dfrac{s}{4}\right)a$
Total Area Of Polygon	$Total\ Area = 2n \cdot A$ $Total\ Area = 2n \cdot \left(\dfrac{s}{4}\right)a$ $Total\ Area = \left(\dfrac{n \cdot s}{2}\right)a$

Two Points

Two Points	$A\,(\,a_1,\,a_2\,)$ and $B\,(\,b_1,\,b_2\,)$
MIDPOINT between 2 points	$M\ =\ $(Average X, Average Y) $M\ =\ \left(\dfrac{a_1+b_1}{2}\,,\ \dfrac{a_2+b_2}{2}\right)$
DISTANCE between 2 points	$D\ =\ \sqrt{(\Delta x)^2\ +\ (\Delta y)^2}$ $\Delta =$ Difference $D\ =\ \sqrt{(b_1\ -\ a_1)^2\ +\ (b_2\ -\ a_2)^2}$
SLOPE of a line between 2 points	Slope $=\ \dfrac{Rise}{Run}\ =\ \dfrac{Change\ in\ y}{Change\ in\ x}$ Slope $=\ \dfrac{\Delta y}{\Delta x}\ =\ \dfrac{(b_2-a_2)}{(b_1-a_1)}$
EQUATION of a line through 2 points	Format: $y\ =\ mx\ +\ b$ Where: $m\ =$ Slope. $b\ =$ y-Intercept

Linear Equations

Two Points	(x_1, y_1) and (x_2, y_2)
SLOPE of a line between 2 points	$\text{Slope} = m = \dfrac{Rise}{Run} = \dfrac{Change\ in\ y}{Change\ in\ x}$ $m = \dfrac{\Delta y}{\Delta x} = \dfrac{(y_2 - y_1)}{(x_2 - x_1)}$
Slope-Intercept Form	$y = mx + b$
Point-Slope Form	$(\Delta y) = m(\Delta x)$ $(y - y_1) = m(x - x_1)$ or $\quad (y - y_2) = m(x - x_2)$
Standard Form	$Ax + By = C$
Parallel Lines	$m_1 = m_2$
Perpendicular Lines	$m_1 = -\dfrac{1}{m_2} \quad , \ m_2 \neq 0$

Quadratic Equations

Standard Form	$f(x) = ax^2 + bx + c$ Zeros: Factor it or use the Quadratic Formula.	Vertex: $\left(\left(\frac{-b}{2a}\right), f\left(\frac{-b}{2a}\right)\right)$ AOS: $x = \frac{-b}{2a}$
Vertex Form	$f(x) = a(x - h)^2 + k$	Vertex: (h, k) AOS: $x = h$
Factored Form	$f(x) = a(x - z_1)(x - z_2)$ Zeros: z_1 and z_2	Vertex: $\left(\left(\frac{z_1 + z_2}{2}\right), f\left(\frac{z_1 + z_2}{2}\right)\right)$ AOS: $x = \frac{z_1 + z_2}{2}$
Far End Behavior	$a > 0$ Positive Happy Parabola	
	$a < 0$ Negative Sad Parabola	
Quadratic Formula	If $ax^2 + bx + c = 0$ Then $x = \dfrac{-b \pm \sqrt{b^2 - 4ac}}{2a}$	
Special Products	$(a + b)^2 = a^2 + 2ab + b^2$	
	$(a - b)^2 = a^2 - 2ab + b^2$	
	$(a + b)(a - b) = a^2 - b^2$ "Difference of Two Squares"	
Binomial Expansion	$(a + b)^n = \sum_{k=0}^{n} {}_nC_k \cdot a^{n-k} \cdot b^k$ Where: ${}_nC_k = \dfrac{n!}{(n - k)!\, k!}$	

AOS = Axis of Symmetry

Geometry Circles

	$m\sphericalangle 1 = m°$
	$m\sphericalangle 1 = \frac{1}{2} m°$
	$m\sphericalangle 1 = \frac{1}{2} m°$
	$m\sphericalangle 1 = \frac{1}{2}(m+n)°$ $ab = cd$
	$m\sphericalangle 1 = \frac{1}{2}(m-n)°$ $qr = st$
	$m\sphericalangle 1 = \frac{1}{2}(m-n)°$ $qr = t^2$
	$m\sphericalangle 1 = 180° - n° = m° - 180°$ $a = b$
	$length\ \overset{\frown}{AB} = \left(\frac{m°}{360°}\right) \cdot 2\pi r$ $m° = m\sphericalangle AOB = m\overset{\frown}{AB}$

Unit Circle

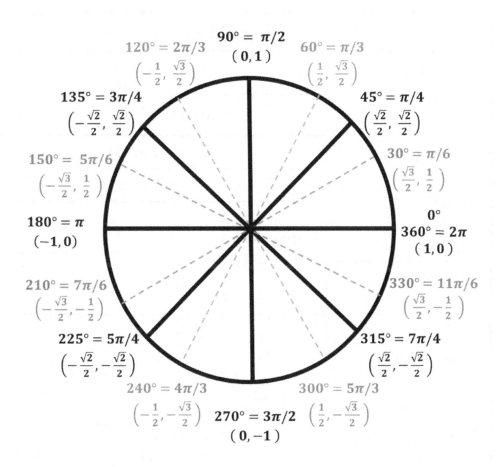

Triangles

<table>
<tr><td colspan="3" align="center">TRIANGLES</td></tr>
<tr>
<td rowspan="4">Right Triangle
</td>
<td colspan="2" align="center">$Cos(\theta) = \dfrac{adjacent}{hypotenuse}$</td>
</tr>
<tr><td colspan="2" align="center">$Sin(\theta) = \dfrac{opposite}{hypotenuse}$</td></tr>
<tr><td colspan="2" align="center">$x = r \cdot Cos(\theta)$</td></tr>
<tr><td colspan="2" align="center">$y = r \cdot Sin(\theta)$</td></tr>
<tr>
<td rowspan="2">Pythagorean Theorem</td>
<td align="center">$x^2 + y^2 = r^2$</td>
</tr>
<tr><td align="center">$sin^2(\theta) + cos^2(\theta) = 1$</td></tr>
<tr>
<td rowspan="3">Any Triangle
</td>
<td>Law of Sines</td>
<td align="center">$\dfrac{a}{sin(A)} = \dfrac{b}{sin(B)} = \dfrac{c}{sin(C)}$</td>
</tr>
<tr>
<td>Law of Cosines</td>
<td align="center">$a^2 = b^2 + c^2 - 2bc \cdot Cos(A)$

Pythagorean Theorem   —   Correction</td>
</tr>
<tr>
<td>Area</td>
<td align="center">$s = $ Semi Perimeter $= \dfrac{a+b+c}{2}$

Area $= \sqrt{s\,(s-a)(s-b)(s-c)}$</td>
</tr>
</table>

Exponents

$$a^m \cdot a^n = a^{(m+n)}$$

$$\frac{a^m}{a^n} = a^{(m-n)}$$

$$\frac{a^n}{a^n} = a^{(n-n)} = a^0 = 1$$

$$(a \cdot b \cdot c)^n = a^n \cdot b^n \cdot c^n$$

$$(a^m)^n = a^{m \cdot n}$$

$$a^{-n} = \frac{1}{a^n}$$

$$\left(\frac{a}{b}\right)^{-n} = \left(\frac{b}{a}\right)^n = \frac{b^n}{a^n}$$

$$\sqrt[n]{a^m} = a^{\left(\frac{m}{n}\right)}$$

$$\sqrt[n]{a^n} = a^{\left(\frac{n}{n}\right)} = a^1 = a$$

Note: If n is an even number $\sqrt[n]{a^n} = |a|$

Factoring

Perfect Square Trinomials	$(a+b)^2 = a^2 + 2ab + b^2$ $(a-b)^2 = a^2 - 2ab + b^2$
Difference of 2 Squares	$a^2 - b^2 = (a-b)(a+b)$
Sum and Diff. of 2 Cubes	$a^3 + b^3 = (a+b)(a^2 - ab + b^2)$ $a^3 - b^3 = (a-b)(a^2 + ab + b^2)$
Distributive Property	$(a+b)(c+d+e)$ $= ac + ad + ae + bc + bd + be$
Quadratic Formula	If: $\quad ax^2 + bx + c = 0$ Then: $\quad x = \dfrac{-b \pm \sqrt{b^2 - 4ac}}{2a}$
Binomial Expansion	$(a+b)^n = \sum_{k=0}^{n} \binom{n}{k} a^{n-k} b^k$
Combination	$\binom{n}{k} = \dfrac{n!}{k!\,(n-k)!}$
Pascal's Triangle	1 1 1 1 2 1 1 3 3 1 1 4 6 4 1 1 5 10 10 5 1

Exponential Functions

	$y = b^x$ Exponential Growth $b > 1$	$y = b^x$ Exponential Decay $b < 1$
Exponential Function	$b > 1 \rightarrow$ Growth $b < 1 \rightarrow$ Decay $a =$ Initial Amt.	$f(x) = ab^x$
Growth of Money $A = Amt.\,after\,t\,yrs.$ $P = Principal$ $r = Yearly\,Rate$ $t = Years$	Compounded Yearly	$A = P(1 + r)^t$
	Compounded n times per year	$A = P\left(1 + \frac{r}{n}\right)^{nt}$
	Compounded Continuously	$A = Pe^{rt}$
Decay of Value $A_0 = Initial\,Value$	Depreciates at a yearly rate of r	$A = A_0(1 - r)^t$
	Depreciates by half every h years	$A = A_0 \left(\frac{1}{2}\right)^{t/h}$

Logs

	$y = log_b(x)$ $b > 1$	$y = log_b(x)$ $b < 1$

Definition of a LOG	$log_b(A) = n \quad \Leftrightarrow \quad b^n = A$
Reference Points $(1,0)$ and $(b,1)$	$log_b(1) = 0 \quad \Leftrightarrow \quad b^0 = 1$
	$log_b(b) = 1 \quad \Leftrightarrow \quad b^1 = b$
Conventions	$log_{10}(A) = log(A)$ $log_e(A) = ln(A)$
Product Property	$log(A) + log(B) = log(AB)$
Quotient Property	$log(A) - log(B) = log\left(\frac{A}{B}\right)$
Power Property	$log(A^n) = n \cdot log(A)$
Reciprocal Property	$log\left(\frac{1}{A}\right) = log(A^{-1}) = -log(A)$
Inverse Properties	$log_b(b^x) = x$
	$b^{log_b(x)} = x$
Change of Base	$log_b(A) = \frac{log_c(A)}{log_c(b)} = \frac{log(A)}{log(b)} = \frac{ln(A)}{ln(b)}$

Parent Functions

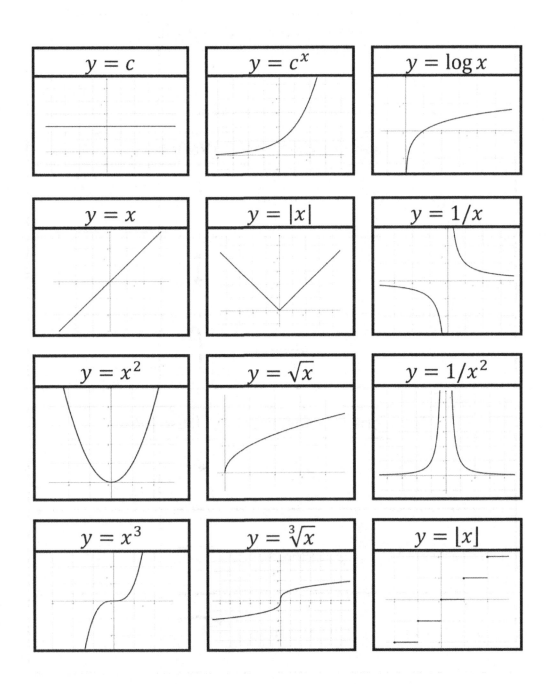

Translations

Parent Function (Example)	
$f(x) = x^2$	

Vertical Translations (Very Nice)		
$y = f(x) + a$	Shift Up	↑
$y = a * f(x)$	Stretch	↕
$y = -f(x)$	V. Rotation	

Horizontal Translations (Horrible)		
$y = f(x + a)$	Shift Left	←
$y = f(a * x)$	Compression	→ ←
$y = f(-x)$	H. Rotation	

Graphing Polynomials

Example	$y = a(x - z_1)(x - z_2)^2(x - z_3)^3(x - z_4)^4$ $y = a \cdot x^{10} \quad \dots \quad \dots$
Degree	$D = 1 + 2 + 3 + 4 = 10$

<table>
<tr><th rowspan="3">Far-End
Behavior</th><th></th><th>Similar to</th><th>Positive a</th><th>Negative a</th></tr>
<tr><td>Odd D</td><td>1st Degree</td><td>↙ ↗</td><td>↖ ↘</td></tr>
<tr><td>Even D</td><td>2nd Degree</td><td>↖ ↗</td><td>↙ ↘</td></tr>
</table>

Zeros and Multiplicity	$z_1, \; m1 \qquad z_2, \; m2 \qquad z_3, \; m3 \qquad z_4, \; m4$
Behavior at Zeros	Odd Multiplicity → Through Even Multiplicity → Bounce

Graphing Example Even Degree Positive a	

Y-Intercept	$x = 0 \;\rightarrow\; y = a(-z_1)(-z_2)^2(-z_3)^3(-z_4)^4$

Graphing Rational Functions

For Rational Functions in the form:

$$R(x) \; = \; \frac{f(x)}{g(x)} \; = \; \frac{a\,(x - n_1)\cdot(x - n_2)\;\text{...}}{b\,(x - d_1)\cdot(x - d_2)\;\text{...}}$$

With: D_n , D_d = Degrees of numerator and denominator

Asymptote	#	Look for this	Form	Does $R(x)$ Cross?
Vertical (VA)	Can be many	Values that make denominator $= 0$	$x = d_1$ $x = d_2$ $x = d_3$ $\ldots$	NEVER crosses. Even mult. → Both up or down. Odd mult. → One up, One down
Horizontal (HA)	One at most	$D_n < D_d$ (bottom heavy)	$y = 0$	May cross. Check: $R(x) = HA$
		$D_n = D_d$	$y = \frac{a}{b}$	
Slant (SA)	One at most	$D_n = D_d + 1$ (a little top-heavy) $f(x) \div g(x)$ Ignore remainder.	$y = mx + b$	May cross. Check: $R(x) = SA$
Hole o	Can be many	Holes occur at zeros (z) of factors that cancel.	$(\, z, R(z)\,)$	Break.

Conic Sections

	Parabola	Ellipse	Hyperbola
Graph	Focal Width = **4p**		
Equation	$(4p)y = x^2$ One gets squared. One gets (4p).	$\dfrac{x^2}{a^2} + \dfrac{y^2}{b^2} = 1$	$\dfrac{x^2}{a^2} - \dfrac{y^2}{b^2} = 1$
AOS	$x = 0$	$y = 0,\ x = 0$	$y = 0,\ x = 0$
Vertex	$(0,0)$	$(\pm a, 0)$	$(\pm a, 0)$
Focus / Foci	$(0, p)$	$(\pm c, 0)$	$(\pm c, 0)$
Directrix	$y = -p$	---	---
Major Axis	---	$y = 0$	$y = 0$
Minor Axis	---	$x = 0$	$x = 0$
Center	---	$(0,0)$	$(0,0)$
Pythagorean Relationship	---	$c^2 = a^2 - b^2$ $a > b$	$c^2 = a^2 + b^2$
Eccentricity	---	$e = \dfrac{c}{a} < 1$	$e = \dfrac{c}{a} > 1$
Asymptotes	---	---	$y = \pm \dfrac{b}{a}x$

AOS = Axis of Symmetry

Series

	ARITHMETIC	GEOMETRIC
Description	Common Difference $d = a_{n+1} - a_n$	Common Ratio $r = \dfrac{a_{n+1}}{a_n}$
Sequence Example	$1, 3, 5, 7, 9 \ldots \quad d = 2$	$1, 2, 4, 8, 16 \ldots \quad r = 2$
Series Example	$1 + 3 + 5 + 7 + 9 + \ldots$	$1 + 2 + 4 + 8 + 16 + \ldots$
Recursive Equation	$a_n = a_{n-1} + d, \quad a_1 = \#$	$a_n = a_{n-1}(r), \quad a_1 = \#$
Explicit Equation	$a_n = a_1 + d(n-1)$	$a_n = a_1(r^{n-1})$
Sum of first n terms	$\sum_{i=1}^{n} a_1 + d(i-1)$ $= n \cdot (Average)$ $= n \cdot \left(\dfrac{a_1 + a_n}{2}\right)$	$\sum_{i=1}^{n} a_1(r^{i-1})$ $= \dfrac{a_1 - a_n(r)}{1-r}$
Sum of terms from $n = j$ to $n = k$	$\sum_{i=j}^{k} a_1 + d(i-1)$ $= (k - j + 1) \cdot (Avg.)$ $= (k - j + 1) \cdot \left(\dfrac{a_j + a_k}{2}\right)$	$\sum_{i=j}^{k} a_1(r^{i-1})$ $= \dfrac{a_j - a_k(r)}{1-r}$
Sum of Infinite Series	Not Possible!!!	Only possible if $\lvert r \rvert < 1$ $\sum_{i=1}^{\infty} a_1(r^{i-1})$ $= \dfrac{a_1 - a_{\infty}(r)}{1-r} \nearrow 0$ $= \dfrac{a_1}{1-r}$

Basic Stats

TERM	POPULATION PARAMETER	SAMPLE STATISTIC
Mean (Average)	$\mu = \dfrac{\sum_{i=1}^{N} x_i}{N}$	$\bar{x} = \dfrac{\sum_{i=1}^{n} x_i}{n}$
Variance	$\sigma^2 = \dfrac{\sum_{i=1}^{N} (x_i - \mu)^2}{N}$	$s^2 = \dfrac{\sum_{i=1}^{n} (x_i - \bar{x})^2}{(n-1)}$
Standard Deviation	σ	s
Median	Middle Term or average of middle terms	
Mode	Term(s) Occurring most often	
Z Score	$z_i = \dfrac{x_i - \mu}{\sigma} = \dfrac{x_i - \bar{x}}{s}$	
Normal Distribution	$\mu \pm 1\sigma = 68\%$ $\mu \pm 2\sigma = 95\%$ $\mu \pm 3\sigma = 99.7\%$	99.7% 95% 68% 34.1% 34.1% 13.6% 13.6% 2.1% 2.1% -3σ -2σ $-\sigma$ Mean σ 2σ 3σ

Vectors

Definition	Vector from point $A = (x_1, y_1, z_1)$ to $B = (x_2, y_2, z_2)$ $\overrightarrow{AB} = \langle x_2 - x_1, \ y_2 - y_1, \ z_2 - z_1 \rangle$ A vector has direction and magnitude.
Magnitude	$\|\vec{p}\| = \|\langle a,b,c \rangle\| = \sqrt{a^2 + b^2 + c^2}$
Unit Vector	$\dfrac{\vec{p}}{\|\vec{p}\|} = \dfrac{\langle a, b, c \rangle}{\sqrt{a^2 + b^2 + c^2}}$
Addition	$\vec{p} + \vec{q} = \langle a,b,c \rangle + \langle d,e,f \rangle$ $\qquad = \langle a+d, \ b+e, \ c+f \rangle$
Dot Product	$\vec{p} \cdot \vec{q} = \langle a,b,c \rangle \cdot \langle d,e,f \rangle = ad + be + cf$ $\qquad = \|\vec{p}\|\|\vec{q}\| \cos \theta \quad \rightarrow \quad \cos \theta = \dfrac{\vec{p} \cdot \vec{q}}{\|\vec{p}\|\|\vec{q}\|}$
Cross Product	$\vec{p} \times \vec{q} = \langle a,b,c \rangle \times \langle d,e,f \rangle$ Vector is $\perp$ to both $\vec{p}$ and $\vec{q}$ $\qquad = \begin{vmatrix} i & j & k \\ a & b & c \\ d & e & f \end{vmatrix}$ $\qquad = \begin{vmatrix} b & c \\ e & f \end{vmatrix} i - \begin{vmatrix} a & c \\ d & f \end{vmatrix} j + \begin{vmatrix} a & b \\ d & e \end{vmatrix} k$ $\qquad = \langle (bf - ce), \ -(af - cd), \ (ae - bd) \rangle$ $\|\vec{p} \times \vec{q}\| = \|\vec{p}\|\|\vec{q}\| \sin \theta \qquad$ If zero $\rightarrow \vec{p} \parallel \vec{q}$
Scalar Projection of $\vec{p}$ onto $\vec{q}$	$comp_q \, \vec{p} = \dfrac{\vec{p} \cdot \vec{q}}{\|\vec{q}\|}$
Vector Projection of $\vec{p}$ onto $\vec{q}$	$proj_q \, \vec{p} = \left(\dfrac{\vec{p} \cdot \vec{q}}{\|\vec{q}\|} \right) \dfrac{\vec{q}}{\|\vec{q}\|} = \dfrac{\vec{p} \cdot \vec{q}}{\|\vec{q}\|^2} \, \vec{q}$
Volume of parallelepiped	$Volume = \|\vec{p} \cdot (\vec{q} \times \vec{v})\|$

Complex Numbers

Imaginary Numbers	$i = \sqrt{-1}$ $i^2 = -1$
Complex Numbers	$z = a + bi$ (Rectangular Form) $z = (r, \theta)$ (Polar Form) $z = r(\cos\theta + i\sin\theta)$ $z = r\ cis\ \theta$ $z = r \cdot e^{i\theta}$
Rectangular To Polar	$r = \sqrt{a^2 + b^2}$ $\theta = \tan^{-1}\left(\dfrac{b}{a}\right)$

Polar Curves

Circle	$r = 2$	
Spiral	$r = \left(\frac{1}{4}\right)\theta$	
Rose	$r = 2\sin(5\theta)$	
Limacon	$r = 2 + \sin\theta$ Bean	
	$r = 2 + 2\sin\theta$ Cardioid	
	$r = 1 + 2\sin\theta$	
	$r = 2\sin\theta$ Circle	

PART 2 – EXAMPLES

The following sections contain examples, using the one-page summaries, included in the previous section.

Examples for polynomial division are also included.

Examples: Lines

LINES: Ex. 1		

Given: A right triangle with legs 3 ft. and 4 ft

Find: The perimeter of a right triangle with legs 3 ft. and 4 ft.

Solution:

Make a sketch	
Use Pythagorean Theorem to find hypotenuse length	$3^2 + 4^2 = h^2$ $9 + 16 = h^2$ $25 = h^2$ $\sqrt{25} = \sqrt{h^2}$ $5 = h$
Add all sides to get perimeter	$P = Perimeter$ $P = 3 + 4 + 5$ $P = 12\ feet$

LINES: Ex. 2

Given: A right isosceles triangle with a hypotenuse, $h = 1$

Find: The lengths of the sides (s). Use Pythagorean Theorem

Solution:

Make a sketch	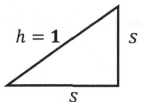 $h = 1$ s s
Use Pythagorean Theorem to find S	$s^2 + s^2 = 1^2$ $2s^2 = 1$ $s^2 = \dfrac{1}{2}$ $s = \sqrt{\dfrac{1}{2}} = \dfrac{\sqrt{1}}{\sqrt{2}} = \dfrac{1}{\sqrt{2}}$ $s = \dfrac{1}{\sqrt{2}} \left(\dfrac{\sqrt{2}}{\sqrt{2}} \right)$ Rationalize the denom. $s = \dfrac{\sqrt{2}}{2}$

Examples: Area

AREA: Ex. 1

Find the area
of the
shaded region.

Use $\pi \approx 3.14$

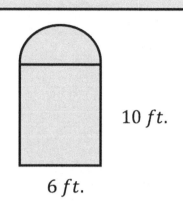

10 ft.

6 ft.

Solution:

$A = Area$

$A = SemiCircle + Rectangle$

$A = \frac{1}{2}[\,Circle\,] + (6)(10)$

$A = \frac{1}{2}[\pi\,r^2] + 60$

$A = \frac{1}{2}[\pi \cdot 3^2] + 60$

$A = \frac{1}{2}[9 \cdot (3.14)] + 60$

$A = \frac{1}{2}[28.26] + 60$

$A = 14.13 + 60$

$A = 74.13\ ft.^2$

Examples: Volume

VOLUME: Ex. 1

Given: A tent has a square base
with base diagonals 6 feet.
The slanted side-edge
of the tent is 5 feet.

Find: The volume of the tent.

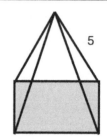

Solution:

Find the area
of the base.

$s^2 = Base\ Area$

$s^2 = 18\ ft^2$

$s^2 + s^2 = 6^2$

$2s^2 = 36$

$s^2 = 18$

$s = \sqrt{18} = \sqrt{9 \cdot 2} = 3\sqrt{2}$

Find the tent
height.

Recall, slanted
side-edge $= 5$

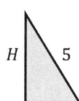

$H^2 + 3^2 = 5^2$

$H^2 = 25 - 9$

$H = \sqrt{16} = 4\ ft.$

3 ← Half the tent diagonal

Now, find the
volume
of the tent.

$V = Volume\ of\ the\ tent$

$V = \left(\frac{1}{3}\right) BH$

$V = \left(\frac{1}{3}\right)(18\ ft^2) \cdot (4\ ft) = 24\ ft^3$

Examples: Regular Polygons

REGULAR POLYGONS: Ex. 1

Given: A regular hexagon has 10 inch sides.
Find: The area.

Solution:

Make a Sketch Hint: A hexagon has 6 sides	
Find the central angle and half the central angle.	$\theta = \frac{360}{6} = 60°$ $\frac{\theta}{2} = 30°$ 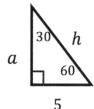
Use the 30 − 60 − 90 triangle properties to find the sides of the right triangle.	$5 = \frac{1}{2} h$ $a = \frac{\sqrt{3}}{2} h$ $(2)5 = (2)\frac{1}{2} h$ $a = \frac{\sqrt{3}}{2}(10)$ $10 = h$ $a = 5\sqrt{3}$
Find the area of the right triangle.	$A = \frac{1}{2}(base)(height)$ $A = \frac{1}{2}(5)(5\sqrt{3}) = \frac{25\sqrt{3}}{2}$
Now, find the total area of the hexagon.	Total Area $= (2 \cdot 6) \cdot \left(\frac{25\sqrt{3}}{2}\right)$ Total Area $= 6 \cdot 25\sqrt{3}$ Total Area $= 150\sqrt{3} \approx 259.8\ ft^2$

REGULAR POLYGONS: Ex. 2

Given: A regular pentagon has 10 inch sides.
Find: The area.

Solution:

Make a Sketch Hint: A pentagon has 5 sides	
Find the central angle and half the central angle.	$\theta = \frac{360}{5} = 72°$ $\frac{\theta}{2} = 36°$ 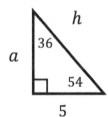
Use the Law of Sines to find apothem. (height of triangle)	$\frac{a}{sin(54)} = \frac{5}{sin(36)}$ $a = \frac{5 \cdot sin(54)}{sin(36)} = 6.88$
Find the area of the right triangle.	$A = \frac{1}{2}\ (base)(height)$ $A = \frac{1}{2}\ (5)(6.88) = 17.2\ ft^2$
Now, find the total area of the pentagon.	Total Area $= (2 \cdot 5)(17.2)$ Total Area $= (10)(17.2)$ Total Area $\approx\ 172\ ft^2$

Examples: Two Points

TWO POINTS: Ex. 1

Problem: Given two points $(-2, 5)$ and $(6, -11)$ Find:
- Slope of the line between them,
- Distance between them,
- Midpoint between them, and
- Equation of a straight line that goes through them.

Solution:

$$Slope = \frac{Change\ in\ y}{Change\ in\ x} = \frac{y_2 - y_1}{x_2 - x_1} = \frac{-11 - 5}{6 - (-2)} = \frac{-16}{8} = -2$$

$$Distance = \sqrt{\Delta x^2 + \Delta y^2}$$

$$Distance = \sqrt{(6 - (-2))^2 + (-11 - 5)^2}$$

$$Distance = \sqrt{(8)^2 + (-16)^2} = \sqrt{320} = 8\sqrt{5}$$

$$Midpoint = \left(\frac{x_1 + x_2}{2}, \frac{y_1 + y_2}{2}\right)$$

$$Midpoint = \left(\frac{-2 + 6}{2}, \frac{5 + (-11)}{2}\right) = \left(\frac{4}{2}, \frac{-6}{2}\right) = (2, -3)$$

$y = mx + b$

$y = (-2)x + b$ Use point $(-2, 5)$ for x and y.

$5 = (-2)(-2) + b$ Solve for b

$5 = 4 + b$ ➔ $1 = b$

The equation is: $y = (-2)x + 1$

Examples: Linear Equations

LINEAR EQUATIONS: Ex. 1

Problem: Given two points: $(2, 5)$ and $(10, 20)$

Find: The equation of line that goes through them in three forms: **Slope-Intercept**, **Point-Slope**, and **Standard Form**.

Solution:

Find the slope	$m \;=\; \dfrac{\Delta y}{\Delta x} \;=\; \dfrac{(20-5)}{(10-2)} \;=\; \dfrac{15}{8}$
Find the y-intercept	$y \;=\; mx \;+\; b$ $(2,5) \;\rightarrow\; 5 \;=\; \left(\dfrac{15}{8}\right)(2) \;+\; b$ $5 \;=\; \left(\dfrac{15}{4}\right) \;+\; b$
	$b \;=\; \left(\dfrac{20}{4}\right) - \left(\dfrac{15}{4}\right) \;=\; \dfrac{5}{4}$
Slope-Intercept Form $y = mx + b$	$y \;=\; \left(\dfrac{15}{8}\right)x \;+\; \left(\dfrac{5}{4}\right)$
Point-Slope Form $(\Delta y) = m(\Delta x)$	Using point $(2, 5)$ $(y - 5) \;=\; \left(\dfrac{15}{8}\right)(x - 2)$
Standard Form $Ax + By = C$	Start with Slope-Intercept form. $y = \left(\dfrac{15}{8}\right)x \;+\; \left(\dfrac{5}{4}\right)$ $8y \;=\; 15x \;+\; 10$ $8y - 15x \;=\; 10$

Examples: Quadratic Equations

QUADRATIC EQUATIONS: Ex. 1

Given $y = x^2 - 6x + 5$

Find the following:
- From given standard form, find the vertex.
- Convert it to vertex then find the vertex.
- Convert it to factored form and find zeros.

Solution:

Standard Form: $y = ax^2 + bx + c$

$$\text{Vertex at:} \quad \left(\frac{-b}{2a}, f\left(\frac{-b}{2a}\right)\right)$$

$y = x^2 - 6x + 5$ $\quad\rightarrow\quad$ Here, $\dfrac{-b}{2a} = \dfrac{6}{2(1)} = 3$

$f(3) = (3)^2 - 6(3) + 5 = -4$ $\quad\rightarrow\quad$ Vertex at: $(3, -4)$

Vertex Form: $y = a(x - h)^2 + k$ $\qquad$ Vertex at: (h, k)

$y = x^2 - 6x + 5 = (x^2 - 6x) + 5$

$y = (x^2 - 6x + 9) + 5 - 9$

$y = (x - 3)^2 - 4$ $\qquad\qquad\rightarrow$ Vertex at: $(3, -4)$

Factored Form: $y = a(x - z_1)(x - z_2)$

Zeros: z_1 and z_2 $\qquad$ Vertex at: $\left(\frac{z_1 + z_2}{2}, f\left(\frac{z_1 + z_2}{2}\right)\right)$

$y = x^2 - 6x + 5$

$y = (x - 1)(x - 5)$ $\qquad\rightarrow$ Zeros at: $x = 1,\ 5$

$\qquad\qquad\qquad\qquad\qquad$ Vertex at: $\left(\frac{1 + 5}{2}, f(3)\right)$

QUADRATIC EQUATIONS: Ex. 2

Given points $(3, -2)$, $(7, -2)$, $(4,4)$, $(8, -12)$

Find: Equation of the parabola that passes through them.
Write the equation in **vertex form**: $y = a(x - h)^2 + k$
HINT: The x-coordinate of the vertex is on AOS.

Solution:

Notice that the first two points have the same y-value. The axis of symmetry (AOS) goes through the middle of $x = 3$ and $x = 7$.	Midpoint is: $\dfrac{3+7}{2} = 5$ AOS: $x = 5$ ➜ $h = 5$. So: $y = a(x - 5)^2 + k$

Use the other points to find "a" and "k"

$(x, y) = (4,4)$ ➜ $4 = a(4 - 5)^2 + k$ $4 = a(-1)^2 + k$ $4 - a = k$	$(x, y) = (8, -12)$ ➜ $-12 = a(4 - 8)^2 + k$ $-12 = a(-3)^2 + k$ $-12 - 9a = k$
$k = k$ $4 - a = -12 - 9a$ $8a = -16$ $a = -2$	$k = 4 - a$ $k = 4 - (-2)$ $k = 4 + 2$ $k = 6$

Vertex Form: $y = a(x - h)^2 + k$ Vertex at (h, k)
$\qquad\qquad\quad y = -2(x - 5)^2 + 6$ Vertex at $(5, 6)$

QUADRATIC EQUATIONS: Ex. 3

The standard form of a quad. eqn. is: $f(x) = ax^2 + bx + c$

For this function: $f(x) = x^2 + 4x - 5$

Find: Vertex, AOS, intercepts, domain, range, and graph it.

Solution:

Vertex $x = -\dfrac{b}{2a}$ $y = f\left(-\dfrac{b}{2a}\right)$	$x = -\dfrac{4}{2(1)} = -2$ $y = (-2)^2 + 4(-2) - 5$ $y = 4 - 8 - 5 = -9$ Vertex at $(x, y) = (-2, -9)$
Axis of Symmetry (AOS) Vertical line thru vertex	$x = -2$ In the form: $x = n$
Y-Intercept $x = 0$	$f(0) = 0^2 + 4(0) - 5 = -5$ Y-Intercept at: $(x, y) = (0, -5)$
X-Intercept $y = 0$	$0 = x^2 + 4x - 5$ $0 = (x + 5)(x - 1)$ $x = -5, 1$ X-Intercepts at: $(-5, 0), (1, 0)$
Domain: All real nums. Range: $-9 \leq y < \infty$	

QUADRATIC EQUATIONS: Ex. 4

The vertex form of a quadratic eqn. is: $\quad y = a(x - h)^2 + k$

with vertex at (h, k)

For this function: $f(x) = -(x - 3)^2 + 4$

Find: Vertex, AOS, intercepts, domain, range, & graph.

Solution:	
Vertex and AOS	$(3, 4)$ $\qquad$ AOS: $x = 3$
Y-Intercept $x = 0$	$f(0) = -(0 - 3)^2 + 4$ $f(0) = -(9) + 4 = -5$ Y-Intercept: $(x, y) = (0, -5)$
X-Intercept $y = 0$	$0 = -(x - 3)^2 + 4$ $(x - 3)^2 = 4$ $x - 3 = \pm\sqrt{4}$ $x = 3 \pm 2 = 5, 1$ X-Intercepts: $(x, y) = (1, 0), (5, 0)$
Domain: $\quad$ All real nums. Range: $\quad -\infty < y \le 4$	

QUADRATIC EQUATIONS: Ex. 5

A quadratic eqn. in factored form is: $y = a(x - z_1)(x - z_2)$
with zeros at z_1 and z_2

For this function: $f(x) = 2(x + 3)(x - 1)$

Find: Vertex, AOS, intercepts, domain, range, & graph.

Solution:	
X-Intercept $y = 0$	$0 = 2(x + 3)(x - 1) \rightarrow x = -3, 1$ X-Intercepts at: $(-3, 0)$, $(1, 0)$
Y-Intercept $x = 0$	$f(0) = 2(0 + 3)(0 - 1) = -6$ Y-Intercept: $(x, y) = (0, -6)$
Vertex x coord. is between the x-intercepts	$x = \dfrac{-3 + 1}{2} = \dfrac{-2}{2} = -1$ $y = f(-1) = 2(-1 + 3)(-1 - 1)$ $y = 2(2)(-2) = -8$ Vertex at: $(x, y) = (-1, -8)$
AOS is a Vertical line through the vertex	$x = -1$ In the form: $x = n$

Domain:
     All real nums.
 	
 Range:
     $-8 \le y < \infty$
	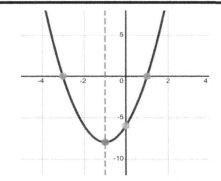

Examples: Geometry Circles

GEOMETRY CIRCLES: Ex. 1

Given: $C = 30^o$ Find: β and θ

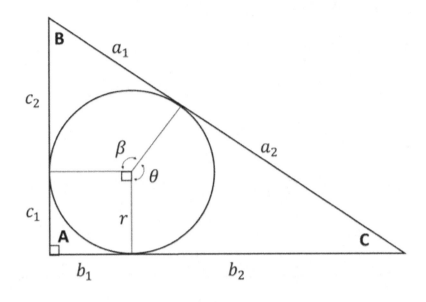

Solution:

Find: θ	$C = 180 - \theta$ $30^o = 180 - \theta$ $\theta = 180 - 30 = 150^o$
Find: β	$B = 90° - C$ $B = 90° - 30 = 60^o$ $60^o = 180 - \beta$ $\beta = 180 - 60$ $\beta = 120^o$

GEOMETRY CIRCLES: Ex. 2

Given: $C = 30^o$ $a_1 = 5$, $r = 3$

Find: a_2 b_1, b_2 c_1, c_2

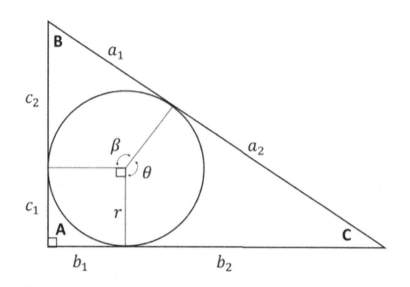

Solution: Let: $a = a_1 + a_2$ $b = b_1 + b_2$ $c = c_1 + c_2$

$A = 90^o$, $B = 60^o$, $C = 30^o$

$c_1 = r = 3$ $c_2 = a_1 = 5$ $c = 5 + 3 = 8$

Law of Sines: $\dfrac{8}{Sin\ 30} = \dfrac{a}{Sin\ 90}$ ➜ $\dfrac{8}{\left(\frac{1}{2}\right)} = \dfrac{a}{(1)}$ ➜ $a = 16$

$a_1 = 5$ (Given) $a_2 = 16 - 5 = 11$

$b_1 = r = 3$ $b_2 = a_2 = 11$

$\dfrac{8}{Sin\ 30} = \dfrac{b}{Sin\ 60}$ ➜ $\dfrac{8}{\left(\frac{1}{2}\right)} = \dfrac{b}{\left(\frac{\sqrt{3}}{2}\right)}$ ➜ $b = 8\sqrt{3} = 13.9$

GEOMETRY CIRCLES: Ex. 3

Given: $a = 10$ and $b_1 = 15$

Find: b Let: $b = b_1 + b_2$

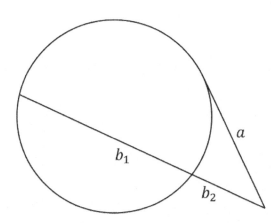

Solution: Let: $b = b_1 + b_2$

$a^2 = b \cdot b_2$

$a^2 = (b_1 + b_2) \cdot b_2$

$(10)^2 = (15 + b_2) \cdot b_2$

$100 = 15 \cdot b_2 + (b_2)^2$ Let $u = b_2$ (below)

$u^2 + 15u - 100 = 0$

$(u - 5)(u + 20) = 0$ ➔ $u = 5, -20$

Use positive solution because distance is positive.

$b_2 = 5$

$b = b_1 + b_2 = 15 + 5 = 20$

Examples: Unit Circle

UNIT CIRCLES: Ex. 1

Find: $sin\left(-\frac{17\pi}{6}\right)$

Solution #1:

- Note the angle is negative so rotate clockwise.

- Also, $-\frac{12\pi}{6} = -2\pi$

- After rotating -2π we are back to original starting position with $\left(-\frac{5\pi}{6}\right)$ more to go.

- And $\left(-\frac{5\pi}{6}\right)$ is just a little short of $\left(-\frac{6\pi}{6}\right)$.

- So, the angle is co-terminal with $\left(\frac{7\pi}{6}\right)$.

- Therefore, $sin\left(-\frac{17\pi}{6}\right) = sin\left(\frac{7\pi}{6}\right) = -\frac{1}{2}$

Solution #2:

- Note that $-\frac{18\pi}{6} = -3\pi$ which is co-terminal with π

- We want $-\frac{17\pi}{6}$ which is a little short of $-\frac{18\pi}{6}$

- So our angle is $\frac{\pi}{6}$ past π in the positive direction.

- So our angle is $\frac{6\pi}{6} + \frac{\pi}{6} = \frac{7\pi}{6}$

- Therefore, $sin\left(-\frac{17\pi}{6}\right) = sin\left(\frac{7\pi}{6}\right) = -\frac{1}{2}$

UNIT CIRCLES: Ex. 2

Given: $\cos\left(\dfrac{x}{2}\right) = -\dfrac{1}{2}$ and $0 \leq x < 2\pi$ Find: x

Solution:

$$\cos\left(\frac{x}{2}\right) = -\frac{1}{2}$$

$$\left(\frac{x}{2}\right) = \cos^{-1}\left(-\frac{1}{2}\right)$$

$$\left(\frac{x}{2}\right) = \frac{2\pi}{3}, \frac{4\pi}{3}$$

There are many places
On the Unit Circle

Where the cosine is $-\dfrac{1}{2}$

$$\left(\frac{x}{2}\right) = \frac{2\pi}{3} + 2k\pi, \ \frac{4\pi}{3} + 2k\pi$$

$$x = \frac{4\pi}{3} + 4k\pi, \ \frac{8\pi}{3} + 4k\pi$$

Recall: $0 \leq x < 2\pi$

So, Just use $k = 0$

$$x = \frac{4\pi}{3}$$

$$x = \frac{8\pi}{3}$$

Final Solution:

$$x = \left\{\frac{4\pi}{3}, \frac{8\pi}{3}\right\}$$

UNIT CIRCLES: Ex. 3

Given: $\sin(2x) = -\dfrac{1}{2}$ and $0 \le x < 2\pi$ Find: x

Solution:

$\sin(2x) = -\dfrac{1}{2}$

$2x = \sin^{-1}\left(-\dfrac{1}{2}\right)$

$2x = \dfrac{7\pi}{6}, \dfrac{11\pi}{6}$

There are many places on the Unit Circle where the sine is $-\dfrac{1}{2}$

$2x = \dfrac{7\pi}{6} + 2k\pi, \dfrac{11\pi}{6} + 2k\pi$

$x = \dfrac{7\pi}{12} + k\pi, \dfrac{11\pi}{12} + k\pi$

Recall: $0 \le x < 2\pi$

So, just use $k = 0, 1$

$x = \dfrac{7\pi}{12}$

$x = \dfrac{7\pi}{12} + \pi = \dfrac{19}{12}\pi$

$x = \dfrac{11\pi}{12}$

$x = \dfrac{11\pi}{12} + \pi = \dfrac{23}{12}\pi$

Final solution:

$x = \left\{ \dfrac{7\pi}{12}, \dfrac{19\pi}{12}, \dfrac{11\pi}{12}, \dfrac{23\pi}{12} \right\}$

UNIT CIRCLES: Ex. 4

Given: $\tan(2x) = 1$ and $0 \le x < 2\pi$ Find: x

Solution:

$\tan(2x) = 1$

$2x = \tan^{-1}(1)$

$2x = \dfrac{\pi}{4}, \dfrac{5\pi}{4}$

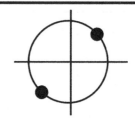

There are many places on the Unit Circle where the tangent is a positive 1.

$2x = \dfrac{\pi}{4} + k\pi, \quad \dfrac{5\pi}{4} + k\pi$

Period for tangent is π radians.

$2x = \dfrac{\pi}{4} + k\pi$ Simplify.

$x = \dfrac{\pi}{8} + \dfrac{k\pi}{2}$ Solve for x

Recall: $0 \le x < 2\pi$ So, just try $k = 0, 1, 2, 3$

$x = \dfrac{\pi}{8}$

$x = \dfrac{\pi}{8} + \dfrac{\pi}{2} \quad = \quad \dfrac{\pi}{8} + \dfrac{4\pi}{8} \quad = \quad \dfrac{5}{8}\pi$

$x = \dfrac{\pi}{8} + \dfrac{2\pi}{2} \quad = \quad \dfrac{\pi}{8} + \dfrac{8\pi}{8} \quad = \quad \dfrac{9}{8}\pi$

$x = \dfrac{\pi}{8} + \dfrac{3\pi}{2} \quad = \quad \dfrac{\pi}{8} + \dfrac{12\pi}{8} \quad = \quad \dfrac{13}{8}\pi$

Final solution: $x = \left\{ \dfrac{\pi}{8}, \dfrac{5\pi}{8}, \dfrac{9\pi}{8}, \dfrac{13\pi}{8} \right\}$

Examples: Triangles

When given some information about a scalene triangle
(all different sides) the Law of Sines and Law of Cosines may
be used to find more information about the given triangle.

Sometimes the given information may be ambiguous where
0, 1, or 2 different triangles may have the given information.
The following table summarizes fixed vs ambiguous triangles.

Given Information		Possible Triangles
SSS	Side-Side-Side	1
ASA	Angle-Side-Angle	1
SAS	Side-Angle-Side	1
SAA	Side-Angle-Angle	1
ASS	Angle-Side-Side	0, 1, 2 **ambiguous**

The first few examples in this section are for fixed triangles.
The last examples are for the **ambiguous** case (ASS or SSA).

TRIANGLES: Ex. 1

Given: $a = 3$, $b = 10$, $c = 8$

Find: All angles for triangle ABC

Solution:

Start by making a sketch!	
Given 3 sides	SSS $\rightarrow$ Only one triangle is possible
Use Law of Cosines to find one angle.	$a^2 = b^2 + c^2 - bc \cdot cos(A)$ $3^2 = 10^2 + 8^2 - 2(10)(8) \cdot cos(A)$ $9 = 164 - 160 \cdot cos(A)$ $cos(A) = \dfrac{155}{160}$ $A = cos^{-1}\left(\dfrac{155}{160}\right) = 14.4^{\circ}$
Use Law of Sines To find another angle	$\dfrac{sin(B)}{10} = \dfrac{sin(14.4)}{3}$ $sin(B) = \left(\dfrac{10}{3}\right)sin(14.4)$ $B = sin^{-1}\left[\left(\dfrac{10}{3}\right)sin(14.4)\right] = 56^{\circ}$
180 $^{\circ}$ Total	$C = 180 - 14.4 - 56 = 109.6^{\circ}$

TRIANGLES: Ex. 2

Given: $A = 40°$, $B = 80°$ with side $c = 6$ $inches$

Find: Sides a, b and the area of the triangle.

Solution:

Start by making a sketch!	B 80° 6 A 40° 60° C a b

180 ° Total	$C = 180 - 80 - 40 = 60°$
Law of Sines	$$\frac{a}{sin(40)} = \frac{6}{sin(60)}$$ $$a = \frac{6 \cdot sin(40)}{sin(60)} = 4.45$$
Law of Sines	$$\frac{b}{sin(80)} = \frac{6}{sin(60)}$$ $$b = \frac{6 \cdot sin(80)}{sin(60)} = 6.82$$

$$s = \text{Semi Perimeter} = \frac{4.45 + 6.82 + 6}{2} = 8.64$$

$$Area = \sqrt{s(s-a)(s-b)(s-c)}$$

$$Area = \sqrt{8.64(8.64-4.45)(8.64-6.82)(8.64-6)}$$

$$Area = 13.19 \ in^2$$

TRIANGLES: Ex. 3

Given:	Triangle ABC with $A = 20°$
	and sides $c = \#c, \quad a = \#a$
Describe:	Possible number of tringles (Use common sense)

Solution:

Start by making a sketch!	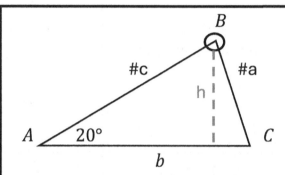

Given Angle-Side-Side (ASS)	0, 1, *or* 2 cases may be possible. Think of angle B as a hinge between the two given sides. This is an AMBIGUOUS case.

# Cases	Conditions	Description
0	$\#a < h$ too short	$\#a$ can't swing down to touch side b No triangle is possible.
1	$\#a = h$ just right!	$\#a$ forms a right angle with side b. Only one triangle is possible.
1	$\#a > \#c$	$\#a$ too long to swing to side c. Only one triangle is possible.
2	$h < \#a < \#c$	$\#a$ can swing outward, or inward

TRIANGLES: Ex. 4

Given: Triangle ABC with $A = 20°$ and sides $c = 10$, $a = ?$

Find: The value for a that will make a right angle at C.
Then, using that value for a, find b.

Solution:

Start by making a sketch!	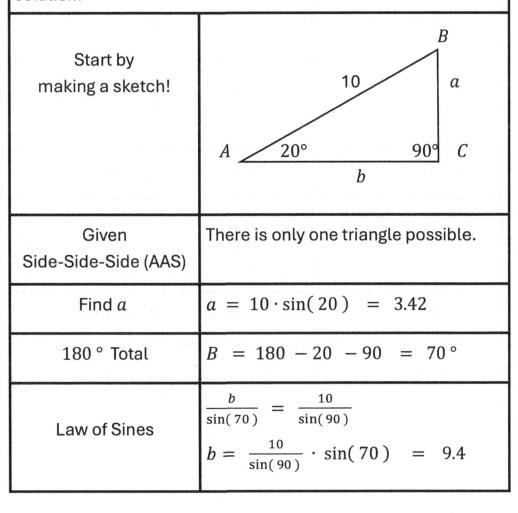
Given Side-Side-Side (AAS)	There is only one triangle possible.
Find a	$a = 10 \cdot \sin(20) = 3.42$
$180°$ Total	$B = 180 - 20 - 90 = 70°$
Law of Sines	$\dfrac{b}{\sin(70)} = \dfrac{10}{\sin(90)}$ $b = \dfrac{10}{\sin(90)} \cdot \sin(70) = 9.4$

TRIANGLES: Ex. 5

Given: Triangle ABC with $A = 20°$ and $c = 10,\ a = 3$

Find: All sides and all angles of the triangle.

Solution:

Start by making a sketch!	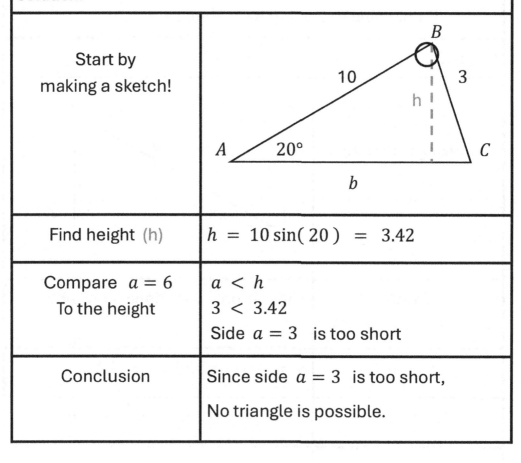
Find height (h)	$h\ =\ 10\sin(\,20\,)\ =\ 3.42$
Compare $a = 6$ To the height	$a\ <\ h$ $3\ <\ 3.42$ Side $a = 3$ is too short
Conclusion	Since side $a = 3$ is too short, No triangle is possible.

TRIANGLES: Ex. 6a

Given: Triangle ABC with $A = 20°$ and $c = 10, a = 16$

Find: All sides and all angles of the triangle.

Solution:

Start by making a sketch!	
Find height (h)	$h = 10\sin(20) = 3.42$
Compare $a = 16$ To the height	$a > h$ $16 > 3.42$
Compare $a = 16$ To side $c = 10$	$a > c$ $16 > 10$
Conclusion	One triangle is possible, Angle B is obtuse & Angle C is acute.

We still need to find Angles C and B and also, side b.

Continued ...

TRIANGLES: Ex. 6b

Given: Triangle ABC with $A = 20°$ and $c = 10, a = 16$

Find: All sides and all angles of the triangle.

Solution:

Start by making a sketch!	

Previously Found	One triangle is possible, Angle B is obtuse & Angle C is acute.
Find angle C Use Law of Sines	$$\frac{sin(C)}{10} = \frac{sin(20)}{16}$$ $$sin(C) = \frac{10 \cdot sin(20)}{16}$$ $$C = sin^{-1}\left(\frac{10 \cdot sin(20)}{16}\right) = 12.34°$$
Find angle B	$B = 180 - 20 - 12.34 = 147.66°$
Find side b Use Law of Sines	$$\frac{b}{sin(147.66)} = \frac{16}{sin(20)}$$ $$b = \frac{16 \cdot sin(147.66)}{sin(20)} = 25.03$$

TRIANGLES: Ex. 7a

Given: Triangle ABC with $A = 20°$ and $c = 10$, $a = 6$

Find: All sides and all angles of the triangle.

Solution:

Start by making a sketch!	*(sketch: triangle with vertices A, B, C; side 10 from A to B, 20° angle at A, side 6 from B to C, height h, base b)*
Find height (h)	$h = 10\sin(20) = 3.14$
Compare $a = 16$ To the height	$a > h$ $6 > 3.14$
Compare $a = 16$ To side $c = 10$	$a < c$ $6 > 10$
Conclusion	Two triangles are possible. Case 1: B is obtuse & C is acute. Case 2: B is acute & C is obtuse

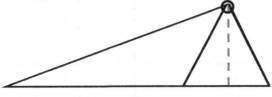

TRIANGLES: Ex. 7b

Given: Triangle ABC with $A = 20°$ and $c = 10$, $a = 6$

Find: All sides and all angles of the triangle.

Solution:

Start by making a sketch!	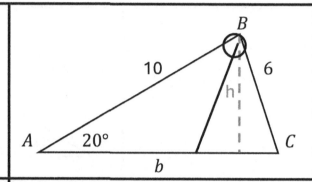

Previously Found	Two triangles are possible.
	Case 1: B is obtuse & C is acute.
	Case 2: B is acute & C is obtuse

Find Angle C: Two possibilities

$$\frac{sin(C)}{10} = \frac{sin(20)}{6}$$

$$C = sin^{-1}\left[\frac{10}{6} sin(20)\right] = 34.75°, \ 145.25°$$

Case 1:	Case 2:
$C = 34.75°$ (acute)	$C = 145.25$ (obtuse)
$B = 180 - 20 - 34.75$	$B = 180 - 20 - 145.25$
$B = 125.25°$	$B = 14.75°$
$\dfrac{b}{sin(125.25)} = \dfrac{6}{sin(20)}$	$\dfrac{b}{sin(14.75)} = \dfrac{6}{sin(20)}$
$b = 14.33$	$b = 4.47$

TRIANGLES: Ex. 8a

Given: Triangle ABC with $A = 30°$ and sides $c = 10$, $a = 5$

Find: All sides and all angles of the triangle.

Solution:

Start by making a sketch!	
Find height (h)	$h = 10\sin(30) = 5$
Compare $a = 5$ To the height	$a = h$ $5 = 5 \quad \rightarrow$ It's a right triangle
Conclusion	One triangles is possible. Because it's a right triangle.
Angles A, B, C	$A = 30 \qquad B = 60 \qquad C = 90$
Sides a, b, c	$a = 5 \qquad c = 10$ $b = 10\cos(30) = 8.66$

Examples: Exponents

EXPONENTS Ex. 1

Simplify the given exponential expressions.

Expression	Simplified Expression
$(a^{-2}b)^2(ab)^{-2}$	$= a^{-4}b^2 \cdot a^{-2}b^{-2}$ $= a^{-6}b^0 \quad = \quad \dfrac{1}{a^6}$
$\left(\dfrac{a^{-2}b}{a^3b^{-4}}\right)^3$	$= \left(\dfrac{bb^4}{a^3a^2}\right)^3 = \left(\dfrac{b^5}{a^5}\right)^3 = \dfrac{b^{15}}{a^{15}}$
$\left(\dfrac{x^{-3}y^{-4}}{x^{-2}y}\right)^{-3}$	$= \left(\dfrac{x^2}{x^3yy^4}\right)^{-3} = \left(\dfrac{1}{xy^5}\right)^{-3}$ $= x^3y^{15}$
$\dfrac{(-3a^2b^6)^2}{(-2ab^7)^3}$	$= -\dfrac{3^2\,a^4\,b^{12}}{2^3\,a^3\,b^{21}} \quad = \quad -\dfrac{9a}{8b^9}$
$\dfrac{16z^{\frac{4}{5}}}{12z^{\frac{1}{5}}}$	$= \left(\dfrac{4}{3}\right)z^{\frac{3}{5}}$
$\left(-5x^{\frac{1}{3}}\right)\left(-6x^{\frac{1}{2}}\right)$	$= 30x^{\left(\frac{1}{3}+\frac{1}{2}\right)} = 30x^{\frac{5}{6}}$

Examples: Factoring

FACTORING Ex. 1		
Factor the trinomials: $x^2 + 5x + 6$ and $x^2 - 5x + 6$		
Solution		
Trinomial	Factoring Steps	Notes
$x^2 + 5x + 6$	$(x\quad)(x\quad)$	$x \cdot x = x^2$
	$(x\quad3)(x\quad2)$	$3 \cdot 2 = 6$
	$(x + 3)(x + 2)$	$3 + 2 = 5$
$x^2 - 5x + 6$	$(x\quad)(x\quad)$	$x \cdot x = x^2$
	$(x\quad3)(x\quad2)$	$3 \cdot 2 = 6$
	$(x - 3)(x - 2)$	$(-3) + (-2) = -5$

$x^2 + 5x + 6 \;=\; (x + 3)(x + 2)$
$x^2 - 5x + 6 \;=\; (x - 3)(x - 2)$

	FACTORING Ex. 2	

Factor the trinomials: $x^2 + x - 6$ and $x^2 - x - 6$

Solution

Trinomial	Factoring Steps	Notes
$x^2 + x - 6$	$(x \quad)(x \quad)$	$x \cdot x = x^2$
	$(x \quad 3)(x \quad 2)$	$3 \cdot 2 = 6$
	$(x + 3)(x - 2)$	$3 - 2 = 1$
$x^2 - x - 6$	$(x \quad)(x \quad)$	$(3) \cdot (-2) = -6$
	$(x \quad)(x \quad)$	$(-3) + (2) = -1$
	$(x - 3)(x + 2)$	$(-3) \cdot (2) = -6$

$x^2 + x + 6 \quad = \quad (x + 3)(x - 2)$
$x^2 - x + 6 \quad = \quad (x - 3)(x \mp 2)$

FACTORING Ex. 3

Factor the trinomials: $2x^2 + 7x + 3$ and $2x^2 + 5x + 3$

Solution

Trinomial	Factoring Steps	Notes
$2x^2 + 7x + 3$	$(2x\quad)(x\quad)$	$2x \cdot x = 2x^2$
	$(2x\quad 1)(x\quad 3)$	$1 \cdot 3 = 3$
	$(2x + 1)(x + 3)$	$6x + 1x = 7x$
		Outer + Inner
$2x^2 + 5x + 3$	$(2x\quad)(x\quad)$	$2x \cdot x = 2x^2$
	$(2x\quad 3)(x\quad 1)$	$3 \cdot 1 = 3$
	$(2x + 3)(x + 1)$	$2x + 3x = 5x$
		Outer + Inner

$$2x^2 + 7x + 3 = (2x + 1)(x + 3)$$

$$2x^2 + 5x + 3 = (2x + 3)(x + 1)$$

FACTORING Ex. 4. (Binomial Expansion)

Find the binomial expansion for: $(x + 2)^3$

Solution

| Binomial Expansion | $(a + b)^n = \binom{n}{k} a^k b^{n-k}$ |
| | $(x + 2)^3 = \sum_0^3 \binom{3}{k} x^k (2)^{n-k}$ |

$(x + 2)^3 =$

$$= \binom{3}{0} x^0 2^3 + \binom{3}{1} x^1 2^2 + \binom{3}{2} x^2 2^3 + \binom{3}{3} x^0 2^3$$

$$= (1) x^0 2^3 + (3) x^1 2^2 + (3) x^2 2^1 + (1) x^3 2^0$$

$$= 8 + 12x + 6x^2 + x^3$$

FACTORING Ex. 5. (Binomial Expansion)	
For the binomial expansion of: $(x - 3)^9$ Find the coefficient of the term with x^6	
Solution	
Binomial Expansion	$(a + b)^n = \binom{n}{k} a^k b^{n-k}$ $(x - 3)^9 = \sum_0^9 \binom{9}{k} x^k (-3)^{n-k}$
To get term with x^6 Use $k = 6$	Term with x^6 $\binom{9}{6} x^6 (-3)^{9-6}$ $(84) x^6 (-3)^3$ $(84) x^6 (-27)$ $(-2268) x^6$
Coefficient of term with x^6	-2268

Examples: Exponential Functions

EXPONENTIAL FUNCTIONS: Ex. 1

Given $80,000 is invested in a bank for 50 years. Find the value of the investment under the following conditions:

Conditions	Result
$80,000 3% interest, compounded annually, for 50 years.	$A = P(1+r)^t$ $A = 80{,}000(1+.03)^{50}$ $A = 80{,}000(1.03)^{50}$ $A = \$350{,}712.50$
$80,000 3% interest, compounded monthly, for 50 years.	$A = P\left(1+\dfrac{r}{12}\right)^{12 \cdot t}$ $A = 80{,}000\left(1+\dfrac{.03}{12}\right)^{12(50)}$ $A = 80{,}000(1.0025)^{600}$ $A = \$357{,}864.60$
$80,000 3% interest, compounded continuously, for 50 years.	$A = P\,e^{rt}$ $A = (80{,}000)\,e^{(.03)(50)}$ $A = (80{,}000)\,e^{1.5}$ $A = \$358{,}535.10$

EXPONENTIAL FUNCTIONS: Ex. 2

Given: Initially, 50 fish in a pond. Increase to 135 after six months. How many fish in 3 years?

Solve, using 3 different formulas, listed below.

Note: 6 months $= \dfrac{1}{2}$ year

Formula	$A = Amount \quad A_0 = Initial\ Amt.$	
$A = A_0\, b^t$ Use initial info. to find b	$135 = (50)\, b^{\left(\frac{1}{2}\right)}$ $b^{\frac{1}{2}} = \dfrac{135}{50} = 2.7 \quad \rightarrow \quad b = 2.7^2 = 7.29$	
	$A = (50)(7.29)^{(3)} \quad = \quad 19371$	
$A = A_0(1 + r)^t$ Use initial info. to find r	$135 = 50(1 + r)^{\left(\frac{1}{2}\right)}$ $\left(\dfrac{135}{50}\right)^2 = (1 + r) \quad \rightarrow \quad r = 6.29$	
	$A = (50)(1 + 6.29)^3 \quad = \quad 19371$	
$A = A_0\, e^{kt}$ Use initial info. to find k	$135 = 50\, e^{k\left(\frac{1}{2}\right)}$ $e^{\frac{k}{2}} = \dfrac{135}{50} = 2.7$ $\ln\left(e^{\frac{k}{2}}\right) = \ln(2.7)$ $\dfrac{k}{2} = \ln 2.7 \quad \rightarrow \quad k = 1.9865$	
	$A = (50)\, e^{(1.9865)(3)} \quad = \quad 19371$	

EXPONENTIAL FUNCTIONS: Ex. 3

Given: Tom paid $50,000 for a new car in 2020.
The value of the car decreases 5% per year.

Find: When will his car be worth $30,000?
Let $t =$ years after 2020. Round t to the nearest year.

Solution:

Find an equation to represent the depreciation of the car.	$A = A_0 (1 + r)^t$ $A = 50,000 (1 - .05)^t$ $A = 50,000 (.95)^t$
Find how many years it will take for the car to depreciate to $30,000	$A = 50,000 (.95)^t$ $30,000 = 50,000 (.95)^t$ $\dfrac{30,000}{50,000} = (.95)^t$ $\dfrac{3}{5} = (.95)^t$ $\log\left[\dfrac{3}{5}\right] = \log\left[(.95)^t\right]$ $\log(.6) = t \cdot \log(.95)$ $\dfrac{\log(.6)}{\log(.95)} = t$ $9.95891 = t \quad \rightarrow \quad t = 10\ yrs.$
Conclusion	In 10 $yrs.$ after 2020, the car will be worth $30 K$. The year will be 2030.

Examples: Logs

LOGS: Ex. 1

Use logs to solve for x in the following equations.

Equation	Solve for x
$log_2 x = 8$	$2^8 = x$ $x = 256$
$log_3 9 = x$	$log_3(3^2) = x$ $2 \cdot log_3 3 = x$ $2(1) = x$ $2 = x$
$log(2x) + log(6)$ $\quad = log(120)$	$log(12x) = log(120)$ $12x = 120$ $x = 10$
$12^{(x+2)} = 123$	$log\left[12^{(x+2)}\right] = log[123]$ $(x+2) \cdot log[12] = log[123]$ $x + 2 = \dfrac{log(123)}{log(12)}$ $x = 1.937 - 2 = -0.063$

LOGS: Ex. 2. (Exponential Decay)	
A fossilized bone contains 12% of its normal amount of carbon 14. How old is the fossil? Use 5600 years as half-life of carbon 14.	
Solution: Let: A = Amount and A_0 = Initial Amount	
Exponential growth or decay function.	$A = A_0 e^{kt}$ For decay, k is negative
Find the value of k. Note: In 5600 years, the carbon 14 content decreases by half.	$(.50) A_0 = A_0 e^{k \cdot 5600}$ $(.50) = e^{k(5600)}$ $\ln(.50) = k(5600)$ $k = \dfrac{\ln(.50)}{5600} \approx -.000124$
Answer the question. Use the k that we found above.	$(.12) A_0 = A_0 e^{(-.000124)t}$ $(.12) = e^{(-.000124)t}$ $\ln(.12) = (-.000124)t$ $t = \dfrac{\ln(.12)}{-.000124} = 17,099$ yrs.
Conclusion	The fossil is about 17,000 years old.

Examples: Parent Functions

PARENT FUNCTIONS: Ex. 1

Given: The equation of a function. $y = (x + 3)^2 - 5$

Find: The parent function and translations. Then graph it.

Solution:

Identify the parent function. $y = x^2$	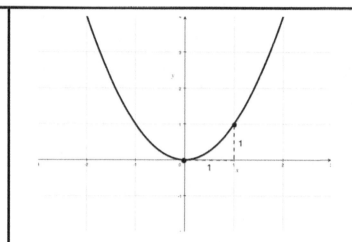
List the horiz. & vertical translations.	• Horizontal shift left by 3. • Vertical shift down by 5. • No stretching or compression.
Graph the function.	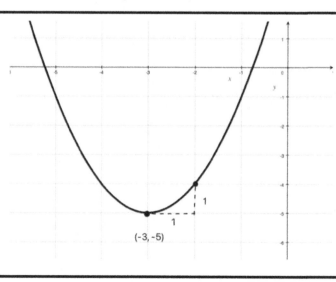

(-3, -5) |

PARENT FUNCTIONS: Ex. 2

Given: The equation of a function. $y = -\sqrt{2x + 6}$

Find: The parent function and translations. Then graph it.

Solution:

Identify the parent function. $y = \sqrt{x}$	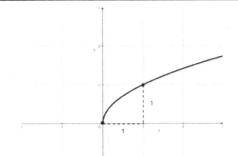
Rewrite equation so coefficient of x is 1.	$y = -\sqrt{2x + 6}$ $y = -\sqrt{2(x + 3)}$
List the horiz. & vertical translations.	• Horizontal shift left by 3. • Horizontal compression by $\frac{1}{2}$ • Vertical rotation over the x-axis.
Graph the function.	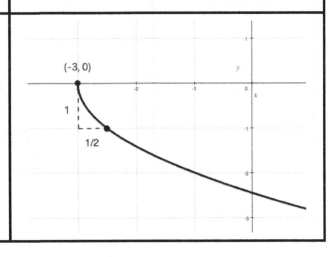

PARENT FUNCTIONS: Ex. 3

Given: The equation of a function. $y = -3|x - 2|$

Find: The parent function and translations. Then graph it.

Solution:

Identify the parent function. $y = |x|$	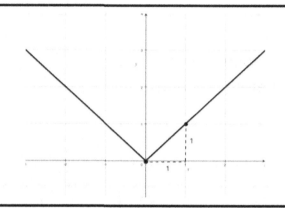
List the horiz. & vertical translations	• Horizontal shift right by 2. • Vertical stretch by 3. • Vertical rotation over the x-axis.
Graph the function.	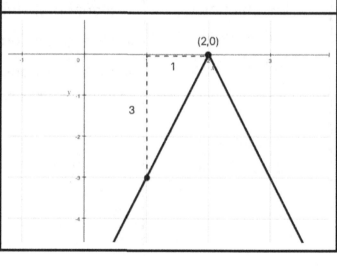

Examples: Translations

TRANSLATIONS: Ex. 1

Given: The function: $f(x) = 3x^2 + 1$ Shift it down by 4 and shift left by 2. Then, stretch it vertically by 2.

Write the eqn. of the transformed function, then graph it.

Solution:

Apply vertical shift up by 4 to the outside.	$[f(x)] - 4 = [3x^2 + 1] - 4$ $f(x) - 4 = 3x^2 - 3$
Apply horizontal shift left by 2 to the (x).	$f(x + 2) - 4 = 3(x + 2)^2 - 3$
Apply vertical stretch by 2 to the outside.	$2[f(x + 2) - 4] = 2[3(x + 2)^2 - 3]$ $2f(x + 2) - 8 = 6(x + 2)^2 - 6$
Compare it to the parent function: $y = x^2$	This is like doing the following translations to the parent function: $y = x^2$ • Vertical stretch by 6 • Horizontal shift left by 2 • And, lastly, vertical shift down by 6

The transformed function is:

$g(x) = 6(x + 2)^2 - 6$

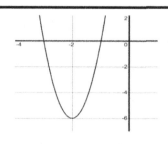

TRANSLATIONS: Ex. 2

Given: The graph of a function. Find: Equation of the function.	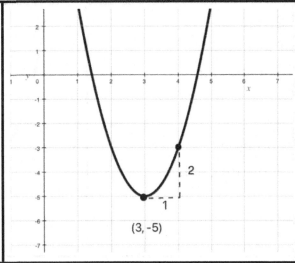 (3, -5)

Solution:

Identify the parent function. $y = x^2$	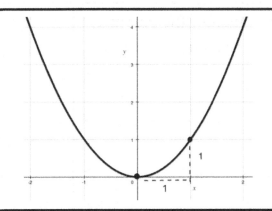
List the horizontal and vertical translations.	• Horizontal shift right by 3. • Vertical shift down by 5. • Vertical stretch by 2.
Write the equation.	$y = 2(x-3)^2 - 5$

TRANSLATIONS: Ex. 3

Given: The graph of a function. Find: Equation of the function.	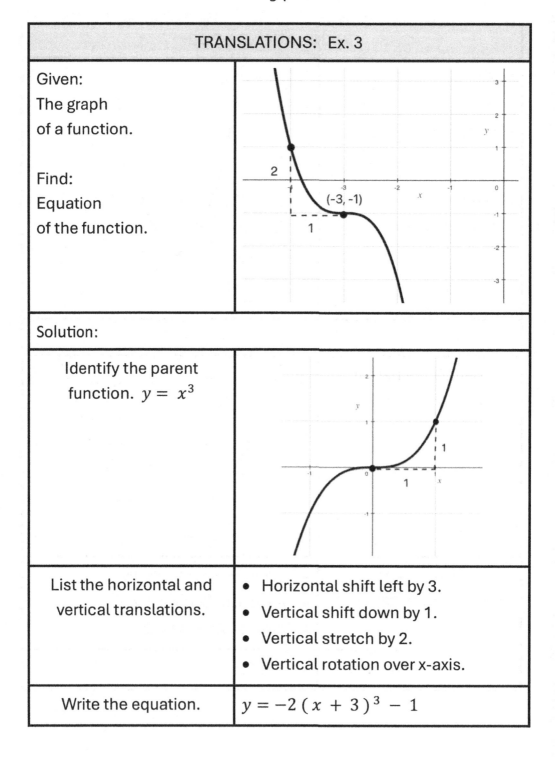

Solution:

Identify the parent function. $y = x^3$	
List the horizontal and vertical translations.	• Horizontal shift left by 3. • Vertical shift down by 1. • Vertical stretch by 2. • Vertical rotation over x-axis.
Write the equation.	$y = -2\,(x + 3)^3 - 1$

TRANSLATIONS: Ex. 4

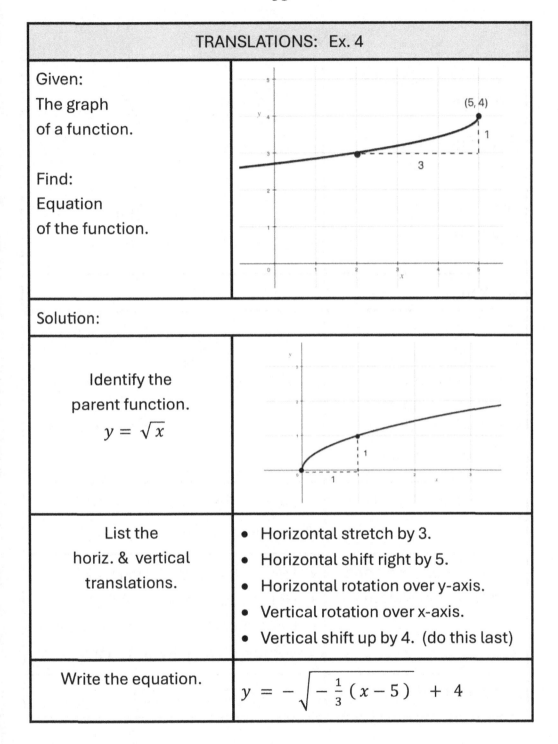

Given:
The graph
of a function.

Find:
Equation
of the function.

(5, 4)

3

1

Solution:

Identify the
parent function.
$$y = \sqrt{x}$$

1

List the horiz. & vertical translations.	• Horizontal stretch by 3. • Horizontal shift right by 5. • Horizontal rotation over y-axis. • Vertical rotation over x-axis. • Vertical shift up by 4. (do this last)
Write the equation.	$$y = -\sqrt{-\tfrac{1}{3}(x-5)} \; + \; 4$$

TRANSLATIONS: Ex. 5

Given: The graph of a function. Find: Equation using the Sine function. Hint: Find beginning & end of sine period.	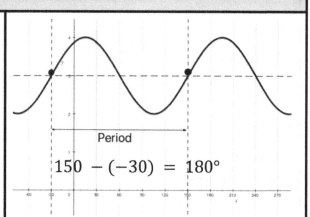 Period $150 - (-30) = 180°$

Solution:

Identify the parent function. $y = \sin x$ Period $= 360°$	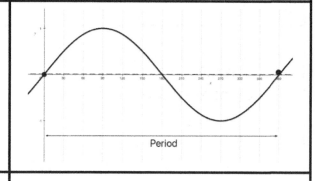 Period
Find the Period.	$Period = 150 - (-30) = 180°$ Given funct. has half a regular period.
List the horizontal and vertical translations.	• Horizontal compression by 2. • Horizontal shift left by 30° • No Vertical Stretch. Amplitude = 1 • Vertical shift up by 3. (do this last)
Write the equation.	$y = \sin(2(x + 30)) + 3$

TRANSLATIONS: Ex. 6	
Given: The graph of a function. Find: Equation using the Cosine function. Hint: Find beginning & end of cosine period.	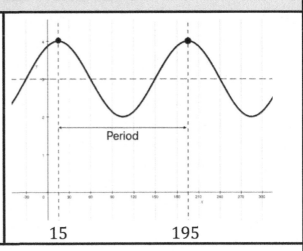 <center>15 195</center>

Solution:	
Identify the parent function. $y = \cos x$ Period $= 360°$	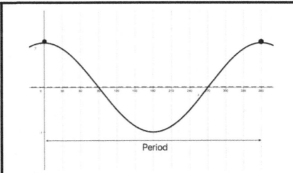
Find the Period.	$Period = 195 - 15 = 180°$ Given funct. has half a regular period.
List the horizontal and vertical translations.	• Horizontal compression by 2. • Horizontal shift right by 15° • Vertical shift up by 3. (do this last)
Write the equation.	$y = \cos(2(x - 15)) + 3$

Examples: Graphing Polynomials

GRAPHING POLYNOMIALS: Ex. 1

Given: $y = 2(x+1)(x-2)(x-3)^2(x-5)^3$

Find: The degree, zeros, multiplicity, far-end behavior and y-intercept. Then sketch it.

Solution:

Degree	$D = 1 + 1 + 2 + 3 = 7$ (odd)
Zeros and multiplicity	$-1, m1$ $2, m1$ $3, m2$ $5, m3$
Far-end behavior	Positive leading coefficient and odd degree. Odd Degree → Like a straight line. → ↙↗
Y-intercept	$y = 2(1)(-2)(-3)^2(-5)^3$ $y = (2)(-2)(9)(-125) = 4500$
Graph	

GRAPHING POLYNOMIALS: Ex. 2

Given: $y = -3 x^4 (x - 5)^2$

Find: The degree, zeros, multiplicity, far-end behavior and y-intercept. Then sketch it.

Solution:

Degree	$D = 4 + 2 = 6$ (Even)
Zeros and multiplicity	0, $m4$ 5, $m2$ Recall, graph bounces at zeros with even multiplicity
Far-end behavior	Negative leading coefficient and even degree. Even Degree → Like parabola → ↙↘
Y-intercept $x = 0$	$y = -3 (0)^4(-5)^2$ $y = 0$
Graph	

Examples: Graphing Rational Functions

GRAPHING RATIONAL FUNCTIONS: Ex. 1

Given: $y = \dfrac{2(x - 1)(x + 2)(x - 4)}{3(x - 1)(x + 3)(x - 6)} = \dfrac{2(x + 2)(x - 4)}{3(x + 3)(x - 6)}$

Find: All asymptotes, holes, and zeros. Then, graph them.

Solution:

Vertical Asymptotes	$x = -3,\ 6$
Horizontal Asymptotes	$y = \dfrac{2}{3} \approx 0.7$
Slant Asymptotes	None. Only if degree of numerator is greater than the denominator by one.
Holes	Hole at $x = 1$ ➔ $(1,\ f(1)) \approx (1,\ 0.3)$
X-intercept	$y = 0$ ➔ $x = -2,\ 4$
Y-Intercept	$x = 0$ ➔ $y = \dfrac{2(2)(-4)}{3(3)(-6)} = \dfrac{8}{27} \approx 0.3$
Graph	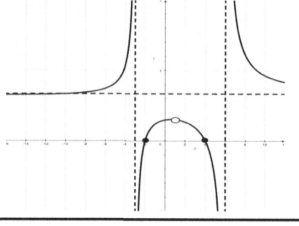

GRAPHING RATIONAL FUNCTIONS: Ex. 2

Given: $y = \dfrac{6x^2 + 7}{x + 1}$

Find: All asymptotes, holes, and zeros. Then, graph them.

Solution:

Vertical Asymptotes	$x = -1$ Multiplicity $= 1$
Horizontal Asymptotes	None.
Slant Asymptotes	$\dfrac{6x^2 + 7}{x + 1} = 6x - 6 + \dfrac{13}{x + 1}$ → $y = 6x - 6$
Holes	None.
X-intercept	$y = 0$ → $x = \pm\sqrt{-\dfrac{7}{6}}$ No real zeros.
Y-Intercept	$x = 0$ → $y = \dfrac{7}{1} = 7$ Y-Intercept at $(0, 7)$
Graph	

GRAPHING RATIONAL FUNCTIONS: Ex. 3

Given: $y = \dfrac{1}{x}$

Find: All asymptotes, holes, and zeros. Then, graph it.

Solution:

Vertical Asymptotes	$x = 0$ Multiplicity $= 1$ ODD
Horizontal Asymptotes	$y = 0$ (Bottom Heavy)
Slant Asymptotes	None.
Holes	None.
X-intercept	None.
Y-Intercept	None. Because $x \neq 0$
Graph	

GRAPHING RATIONAL FUNCTIONS: Ex. 4

Given: $y = \dfrac{1}{x^2}$

Find: All asymptotes, holes, and zeros. Then, graph it.

Solution:

Vertical Asymptotes	$x = 0$ Multiplicity $= 2$ EVEN
Horizontal Asymptotes	$y = 0$ (Bottom Heavy)
Slant Asymptotes	None.
Holes	None.
X-intercept	None.
Y-Intercept	None. Because $x \neq 0$
Graph	

GRAPHING RATIONAL FUNCTIONS: Ex. 5

Given: $y = \dfrac{x^2 - 4}{x^2 - 4x} = \dfrac{(x-2)(x+2)}{x(x-4)}$

Find: All asymptotes, holes, and zeros. Then, graph it.

Solution:

VA	$x = 0, m1$ and $x = 4, m1$
HA	$y = 1$ (Far-End Behavior)
X-Intercept	$y = 0$ → $x = \pm 2$
Y-Intercept	None. Because $x \neq 0$

Does function cross HA?	Set function = HA Solve for x. $\dfrac{x^2 - 4}{x^2 - 4x} = 1$	$4 = 4x$ $x = 1$ Crosses at $(1,1)$

Graph	

GRAPHING RATIONAL FUNCTIONS: Ex. 6		
$y = \dfrac{5 \cdot (x+1)(x+2)(x-2)(x-4)}{2 \cdot (x+1)(x+3)(x-3)(x-6)} = \dfrac{5 \cdot (x+2)(x-2)(x-4)}{2 \cdot (x+3)(x-3)(x-6)}$ Find: All asymptotes, holes, and zeros. Then, graph it.		
Solution:		
VA	$x = -3, 3, 6$ All have $m1$ All ODD	
HA	$y = \dfrac{5}{2} = 2.5$ (Far-End Behavior)	
X-Intercept	$y = 0 \;\rightarrow\; x = -2, 2, 4$	
Y-Intercept	$x = 0 \;\rightarrow\; y = \dfrac{5 \cdot (2)(-2)(-4)}{2 \cdot (3)(-3)(-6)} \approx 0.7$	
Holes	Hole at $x = -1 \;\rightarrow\; (-1, f(-1)) = (-1, 0.67)$	
Does function cross HA?	Set function = HA Solve for x. $\dfrac{5 \cdot (x+2)(x-2)(x-4)}{2 \cdot (x+3)(x-3)(x-6)} = \dfrac{5}{2}$	Function crosses in two places; at $x = -5.6, \; 3.3$
Graph	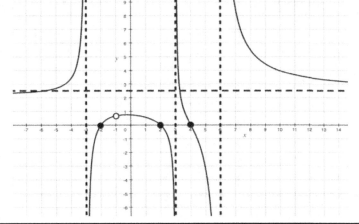	

GRAPHING RATIONAL FUNCTIONS: Ex. 7

Given: $y = \dfrac{x^3 + 8}{(x+1)(x-2)(x-2)}$

Find: All asymptotes, holes, and zeros. Then, graph it.

Solution:

VA	$x = -1, m1 \qquad x = 2, m2$	
HA	$y = 1$ (Far-End Behavior)	
X-Intercept	$y = 0 \quad \rightarrow \quad x = -2$	
Y-Intercept	$x = 0 \quad \rightarrow \quad y = \dfrac{8}{(1)(-2)(-2)} = \dfrac{8}{4} = 2$	
Does function cross HA?	Set function = HA Solve for x. $\dfrac{x^3 + 8}{(x+1)(x-2)(x-2)} = 1$	No real solution. Function does NOT cross HA.
Graph		

GRAPHING RATIONAL FUNCTIONS: Ex. 8

Given: $y = \dfrac{3x^2}{x^2 + 1}$

Find: All asymptotes, holes, and zeros. Then, graph it.

Solution:

VA	None.
HA	$y = \dfrac{3}{1} = 3$ (Far-End Behavior)
X-Intercept	$y = 0$ ➜ $x = 0$ ➜ Point $(0,0)$
Y-Intercept	$x = 0$ ➜ $y = \dfrac{0}{0 + 1} = 0$

Does function cross HA?	Set function = HA Then solve for x $\dfrac{3x^2}{x^2 + 1} = 3$	$x^2 = x^2 + 1$ Not Possible. No Crossings.
Graph		

GRAPHING RATIONAL FUNCTIONS: Ex. 9

Given: $y = \dfrac{3x^2 + 6x}{x^2 + 1} = \dfrac{3x(x+2)}{x^2 + 1}$

Find: All asymptotes, holes, and zeros. Then, graph it.

Solution:

VA	None.
HA	$y = \dfrac{3}{1} = 3$ (Far-End Behavior)
X-Intercept	$y = 0$ → $x = 0, -2$ Points $(0,0), (-2,0)$
Y-Intercept	$x = 0$ → $y = \dfrac{0}{0+1} = 0$

Does function cross HA?	Set function = HA Then, solve for x. $\dfrac{3x^2 + 6x}{x^2 + 1} = 3$	$2x = 1$ $x = \dfrac{1}{2}$ Crosses at $x = \dfrac{1}{2}$

Graph Asymptotes and the function.	

GRAPHING RATIONAL FUNCTIONS: Ex. 10		
Given: $y = \dfrac{x^4 + 1}{x^2} \approx x^2$ Find: All asymptotes, holes, and zeros. Then, graph it.		
Solution:		
VA	$x = 0, m2$	
HA	None.	
SA	None.	
X-Intercept	None.	
Y-Intercept	$x \neq 0$ ➜ NO Y-intercept. ➜ None.	
Does function cross HA?	Set function = HA Solve for x. But, There is no HA!!!	NOTHING TO CHECK!
Graph Asymptotes and the function.	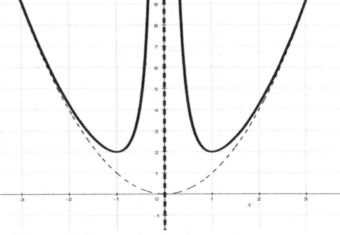	

GRAPHING RATIONAL FUNCTIONS: Ex. 11

Given: $y = \dfrac{x^2}{x+1}$

Find: All asymptotes, holes, and zeros. Then, graph it.

Solution:

VA	$x = -1$	
HA	None.	
SA	$x^2 \div (x+1)$ $x - 1 + \dfrac{1}{x+1}$	SA at $y = x - 1$
X-Intercept	$y = 0$ ➔ $x = 0$ ➔ Point $(0,0)$	
Y-Intercept	$x = 0$ ➔ $y = \dfrac{0}{1} = 0$	
Does function cross SA?	Set function = SA $\dfrac{x^2}{x+1} = x - 1$ $x^2 = (x-1)(x+1)$	$x^2 = x^2 - 1$ Not Possible. No Crossings.
Graph Asymptotes and the function.	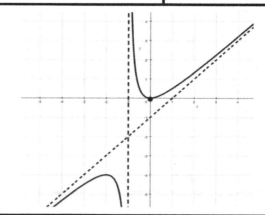	

GRAPHING RATIONAL FUNCTIONS: Ex. 12

Given: $y = \dfrac{x^3}{x^2 + 1}$

Find: All asymptotes, holes, and zeros. Then, graph it.

Solution:

VA	None.	
HA	None.	
SA	$x^3 \div (x^2 + 1) = x - \dfrac{x}{x^2 + 1}$ (Use Polynomial Long Division)	SA at $y = x$
X-Intercept	$y = 0 \quad \rightarrow \quad x = 0 \quad \rightarrow \quad$ Point $(0,0)$	
Y-Intercept	$x = 0 \quad \rightarrow \quad y = \dfrac{0}{0 + 1} = 0$	
Does function cross SA?	Set function = SA $\dfrac{x^3}{x^2 + 1} = x$ $x^3 = x^3 + x$	$0 = x$ Function crosses SA at $x = 0$
Graph Asymptotes and the function.		

GRAPHING RATIONAL FUNCTIONS: Ex. 13

Given: $y = \dfrac{x^3}{x^2 - 1} = \dfrac{x^3}{(x+1)(x-1)}$

Find: All asymptotes, holes, and zeros. Then, graph it.

Solution:

VA	$x = 1, m1$ and $x = -1, m1$
HA	None.

SA	$x^3 \div (x^2 - 1) = x + \dfrac{x}{x^2 - 1}$ (Use Polynomial Long Division)	SA at $y = x$

X-Intercept	$y = 0$ ➔ $x = 0$ ➔ Point $(0,0)$
Y-Intercept	$x = 0$ ➔ $y = \dfrac{0}{0-1} = 0$

Does function cross SA?	Set function = SA $\dfrac{3x^2}{x^2 - 1} = x$ $x^3 = x^3 - x$	$0 = x$ Function crosses SA at $x = 0$

Graph Asymptotes and the function.	

Examples: Conic Sections

CONIC SECTIONS: Ex. 1

Given a quadratic equation in standard form, convert it to a conic section form then graph it. Show the vertex, focus, and directrix. The equation is: $y = \left(\frac{1}{8}\right)x^2 + 4$

Solution:

Find p	One variable gets "4p"
	The other variable gets "squared."
	Since x is squared, the coefficient of y is "4p"
	$(y - 4) = \left(\frac{1}{8}\right)x^2$ $8(y - 4) = x^2$ → $4p = 8$ → $p = 2$

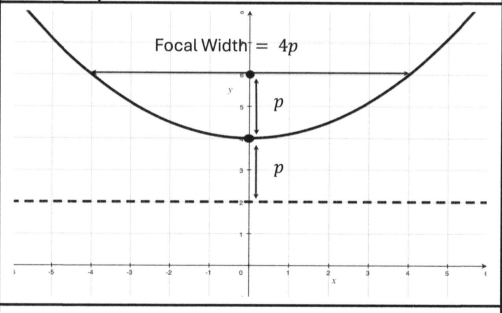

Vertex: $(0, 4)$ Focus: $(0, 6)$ Directrix: $y = 2$

CONIC SECTIONS: Ex. 2

Find the equation of an ellipse and the foci.

Given this equation: $4x^2 + 9y^2 - 48x + 72y + 144 = 0$

Hint: Complete the square twice.

Solution:

$4x^2 + 9y^2 - 48x + 72y + 144 = 0$

$(4x^2 - 48x) + (9y^2 + 72y) = -144$

$4(x^2 - 12x) + 9(y^2 + 8y) = -144$

$4(x^2 - 12x + 36) + 9(y^2 + 8y + 16)$
$$= -144 + 4(36) + 9(16)$$

$4(x - 6)^2 + 9(y + 4)^2 = 144$

$$\frac{(x - 6)^2}{36} + \frac{(y + 4)^2}{16} = 1$$

Center: $(6, -4)$ Major Axis: x

$a = 6$	$c^2 = a^2 - b^2$
$b = 4$	$c = \sqrt{36 - 16} = \sqrt{20} = 2\sqrt{5}$
Foci	$(6 \pm c, -4) = (6 \pm 2\sqrt{5}, -4)$
Major Vertices	$(6 \pm a, -4) = (6 \pm 6, -4)$

CONIC SECTIONS: Ex. 3

Given: An oval, centered at the origin, with $r_1 = 3$ and $r_2 = 4$

Find: The equation of the oval as an Ellipse Conic Section.

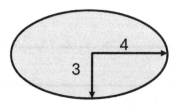

Note: The diagram shows $r_2 = 4$ is on the x axis.

Solution:

$$\frac{x^2}{a^2} + \frac{y^2}{b^2} = 1$$

$$\frac{x^2}{4^2} + \frac{y^2}{3^2} = 1$$

$$\frac{x^2}{16} + \frac{y^2}{9} = 1$$

An easy way to check the solution is to download and use the free, Desmos app on your phone. Or, use the Desmos website from your laptop.

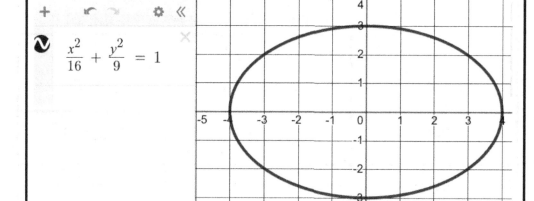

Examples: Series

SERIES: Ex. 1	
Find the sum of the first 100 terms for the arithmetic series: $$1 + 3 + 5 + ...$$	
Solution:	
Find explicit equation for the series	$d = 2$ $a_n = 1 + 2(n-1)$
Find the 100th term.	$a_{100} = 1 + 2(99) = 199$
Find the sum of the first 100 terms.	$S_n = n(Average)$ $S_{100} = (100)\left(\frac{a_1 + a_{100}}{2}\right)$ $S_{100} = (100)\left(\frac{1 + 199}{2}\right) = 10,000$

SERIES: Ex. 2

Find the sum of the first 6 terms and the sum of the infinite geometric series. Given this geometric series:

$$4 - 2 + 1 - \frac{1}{2} \ ...$$

Solution:

Find explicit equation for the series	$r = \frac{-2}{4} = -\frac{1}{2} = -.5$ $a_1 = 4$ $a_n = 4\left(-\frac{1}{2}\right)^{(n-1)}$
Find the 6th term.	$a_6 = 4\left(-\frac{1}{2}\right)^{(6-1)} = -.125$
Find the sum of the first 6 terms.	$S_n = \frac{a_1 - a_n(r)}{1 - r}$ $S_6 = \frac{4 - (-.125)(-.5)}{1 - (-.5)}$ $S_6 = \frac{4 - (.125)(.5)}{1.5} = 2.625$
Find the sum of the infinite geometric series.	$S = \frac{a_1}{1 - r}$ $S = \frac{4}{1 - (-.5)} = \frac{4}{1.5} = 2.6\overline{6}$

SERIES: Ex. 3. (Difficult)		
Given: $\sum_{k=0}^{\infty} ar^{2k} = 5$ and $\sum_{k=1}^{\infty} ar^{2k-1} = 3$		
Find: r Then, find a		

Solution:

Find r	$\sum_{k=0}^{\infty} ar^{2k} = 5$	When $k = 0$ → $a_0 = 5$
	$\sum_{k=1}^{\infty} ar^{2k-1} = 3$	When $k = 1$ → $a_1 = 3$
	$r = \dfrac{a_1}{a_0} = \dfrac{3}{5}$	Therefore: $r = \dfrac{3}{5}$
Find a	$\sum_{k=0}^{\infty} ar^{2k} = 5$	$\sum_{k=1}^{\infty} ar^{2k-1} = 3$
	$a + \sum_{k=1}^{\infty} ar^{2k} = 5$	$\sum_{k=1}^{\infty} ar^{2k} \cdot r^{-1} = 3$
	$\sum_{k=1}^{\infty} ar^{2k} = 5 - a$	$\sum_{k=1}^{\infty} ar^{2k} \cdot \left(\dfrac{5}{3}\right) = 3$
		$\sum_{k=1}^{\infty} ar^{2k} = 3 \cdot \left(\dfrac{3}{5}\right)$
		$\sum_{k=1}^{\infty} ar^{2k} = \dfrac{9}{5}$
	$5 - a = \dfrac{9}{5}$	
	$a = 5 - \dfrac{9}{5} = \dfrac{25}{5} - \dfrac{9}{5} = \dfrac{16}{5}$	

Examples: Basic Stats

BASIC STATS: Ex. 1	
Find the basic statistics, variance, and standard deviation for the following set of sample data: $\{1, 1, 3, 4, 6\}$ Also, find the Z-value for 4 and what it means.	
Solution:	
Mean	$\bar{x} = \dfrac{1 + 1 + 3 + 4 + 6}{5} = \dfrac{16}{5} = 3.2$
Median	Median = Middle term = 3
Mode	Mode = Term occurring most often = 1
Variance	$\dfrac{(1 - 3.2)^2 + (1 - 3.2)^2 + (3 - 3.2)^2 + (4 - 3.2)^2 + (6 - 3.2)^2}{(5 - 1)}$ $\dfrac{4.04 + 4.04 + 0.04 + .64 + 7.84}{4} = 4.15$
Standard Deviation	$s = \sqrt{4.15} = 2.037$
Z-Score for 4	$z = \dfrac{x - \bar{x}}{s} = \dfrac{4 - 3.2}{2.037} = 0.39$
$z = .39$ Meaning	A z-score of .39 means the value is .39 standard deviations above the mean. This corresponds to the 65.2 percentile, using the standard normal tables.

Examples: Vectors

VECTORS: Ex. 1

Given: Points $A(1,2)$, $B(7,10)$, $C(-5,20)$

Find: Vectors $\vec{AB}$, $\vec{AC}$, $\vec{BC}$, $\vec{CB}$ and their magnitudes.

Also find vector $\vec{p} = \vec{AB} - \vec{BC}$ and its magnitude.

Solution:

Vector	Magnitude
$\vec{AB} = \langle 7-1, 10-2 \rangle$ $\vec{AB} = \langle 6,8 \rangle$	$\lvert\vec{AB}\rvert = \sqrt{6^2 + 8^2}$ $\lvert\vec{AB}\rvert = \sqrt{100} = 10$
$\vec{AC} = \langle -5-1, 20-2 \rangle$ $\vec{AC} = \langle -6,18 \rangle$	$\lvert\vec{AC}\rvert = \sqrt{(-6)^2 + 18^2}$ $\lvert\vec{AC}\rvert = \sqrt{360} = 6\sqrt{10}$
$\vec{BC} = \langle -5-7, 20-10 \rangle$ $\vec{BC} = \langle -12,10 \rangle$	$\lvert\vec{BC}\rvert = \sqrt{(-12)^2 + 10^2}$ $\lvert\vec{BC}\rvert = \sqrt{244} = 2\sqrt{61}$
$\vec{CB} = \langle 7+5, 10-20 \rangle$ $\vec{CB} = \langle 12,-10 \rangle$	$\lvert\vec{CB}\rvert = \sqrt{(12)^2 + (-10)^2}$ $\lvert\vec{CB}\rvert = \sqrt{244} = 2\sqrt{61}$
$\vec{p} = \vec{AB} - \vec{BC}$ $\vec{p} = \langle 6,8 \rangle - \langle -12,10 \rangle$ $\vec{p} = \langle 6+12, 8-10 \rangle$ $\vec{p} = \langle 18,-2 \rangle$	$\lvert\vec{p}\rvert = \sqrt{18^2 + (-2)^2}$ $\lvert\vec{p}\rvert = \sqrt{328} = \sqrt{4\cdot82}$ $\lvert\vec{p}\rvert = 2\sqrt{82}$

VECTORS: Ex. 2	
Given: Points $A(1, 2, 3)$, $B(5, 10, 15)$, $C(-1, -2, 0)$	
Find: Vectors $\overrightarrow{AB}$, $\overrightarrow{AC}$, $\overrightarrow{BC}$ and their magnitudes.	
Also find vector $\vec{p} = \overrightarrow{AB} + \overrightarrow{BC}$ and its magnitude.	
Solution:	
Vector	**Magnitude**
$\overrightarrow{AB} = \langle 5-1, 10-2, 15-3 \rangle$ $\overrightarrow{AB} = \langle 4, 8, 12 \rangle$	$\|\overrightarrow{AB}\| = \sqrt{4^2 + 8^2 + 12^2}$ $\|\overrightarrow{AB}\| = \sqrt{224} = 4\sqrt{14}$
$\overrightarrow{AC} =$ $= \langle -1-1, -2-2, 0-3 \rangle$ $= \langle -2, -4, -3 \rangle$	$\|\overrightarrow{AC}\| = \sqrt{2^2 + 4^2 + 3^2}$ $\|\overrightarrow{AC}\| = \sqrt{29}$
$\overrightarrow{BC} =$ $= \langle -1-5, -2-10, 0-15 \rangle$ $= \langle -6, -12, -15 \rangle$	$\|\overrightarrow{BC}\| = \sqrt{6^2 + 12^2 + 15^2}$ $\|\overrightarrow{BC}\| = \sqrt{405} = 9\sqrt{5}$
$\vec{p} = \overrightarrow{AB} + \overrightarrow{BC}$ $\vec{p} =$ $= \langle 4, 8, 12 \rangle + \langle -6, -12, -15 \rangle$ $= \langle -2, -4, -3 \rangle$	$\|\vec{p}\| = \sqrt{2^2 + 4^2 + 3^2}$ $\|\vec{p}\| = \sqrt{29}$

VECTORS: Ex. 3

Given: Three Points $A(1, 2, 3)$, $B(5, 10, 15)$, $C(11, 2, 0)$

Find: Vector $\vec{n}$, perpendicular to the plane defined by the points.

Solution:

$\overrightarrow{AB}$	$\overrightarrow{AB} = \langle 5-1, 10-2, 15-3 \rangle$ $\overrightarrow{AB} = \langle 4, 8, 12 \rangle$
$\overrightarrow{AC}$	$\overrightarrow{AC} = \langle 11-1, 2-2, 0-3 \rangle$ $\overrightarrow{AC} = \langle 10, 0, -3 \rangle$
$\vec{n}$	$\vec{n} = \overrightarrow{AB} \times \overrightarrow{AC}$ $\vec{n} = \begin{vmatrix} i & j & k \\ 4 & 8 & 12 \\ 10 & 0 & -3 \end{vmatrix}$ $\vec{n} = \vec{\imath} \begin{vmatrix} 8 & 12 \\ 0 & -3 \end{vmatrix} - \vec{\jmath} \begin{vmatrix} 4 & 12 \\ 10 & -3 \end{vmatrix} + \vec{k} \begin{vmatrix} 4 & 8 \\ 10 & 0 \end{vmatrix}$ $\vec{n} = (-24-0)\vec{\imath} - (-12-120)\vec{\jmath} + (0-80)\vec{k}$ $\vec{n} = \langle -24, 132, -80 \rangle = 4\langle -6, 33, -20 \rangle$

VECTORS: Ex. 4

Given: Two Vectors $\vec{a} = \langle 4, 8, 12 \rangle$ & $\vec{b} = \langle 10, 0, -3 \rangle$

Find: The angle θ between them.

Solution:

$$\cos(\theta) = \frac{\vec{a} \cdot \vec{b}}{|\vec{a}|\,|\vec{b}|}$$

$$\cos(\theta) = \frac{\langle 4,\ 8,\ 12 \rangle \cdot \langle 10,\ 0,\ -3 \rangle}{\sqrt{4^2 + 8^2 + 12^2}\ \ \sqrt{10^2 + 0^2 + 3^2}}$$

$$\cos(\theta) = \frac{40 + 0 + (-36)}{\sqrt{224}\ \sqrt{109}}$$

$$\cos(\theta) = \frac{4}{\sqrt{224 \cdot 109}} = \frac{4}{\sqrt{24416}} = \frac{4}{4\sqrt{1526}}$$

$$\cos(\theta) \approx \frac{1}{39.064} = 0.02773$$

$$\theta = \cos^{-1}(0.02773)$$

$$\theta = 88.41°$$

VECTORS: Ex. 5

Given: Vectors $\vec{a} = \langle 2, 0, 0 \rangle$, $\vec{b} = \langle 0, 5, 0 \rangle$, $\vec{c} = \langle 0, 0, 8 \rangle$

Find: The volume of the paralepidid.

Note: This is a very simple paralepidid.

It is a rectangular prism with one edge at the origin.

Note: The volume should be: $V = (2)(5)(8) = 80$

Solution:

$Volume = |a \cdot (b \times c)|$

$$\vec{b} \times \vec{c} = \begin{vmatrix} i & j & k \\ 0 & 5 & 0 \\ 0 & 0 & 8 \end{vmatrix}$$

$$= \vec{i} \begin{vmatrix} 5 & 0 \\ 0 & 8 \end{vmatrix} - \vec{j} \begin{vmatrix} 0 & 0 \\ 0 & 8 \end{vmatrix} + \vec{k} \begin{vmatrix} 0 & 5 \\ 0 & 0 \end{vmatrix}$$

$$= 40\,\vec{i} - (0)\,\vec{j} + (0)\,\vec{k}$$

$$= \langle 40, 0, 0 \rangle$$

$$\vec{a} \cdot (\vec{b} \times \vec{c}) = \langle 2, 0, 0 \rangle \cdot \langle 40, 0, 0 \rangle$$

$$= 80 + 0 + 0 = 80$$

$Volume = |\vec{a} \cdot (\vec{b} \times \vec{c})|$

$$= |80| = 80 \qquad \text{As expected.}$$

VECTORS: Ex. 6

Given: Three Vectors:

$$\vec{a} = \langle 1, 2, 3 \rangle, \quad \vec{b} = \langle -1, 1, 2 \rangle, \quad \vec{c} = \langle 2, 1, 4 \rangle$$

Find: The volume of the paralepidid.

Solution:

$$Volume = |a \cdot (b \times c)|$$

$$\vec{b} \times \vec{c} = \begin{vmatrix} i & j & k \\ -1 & 1 & 2 \\ 2 & 1 & 4 \end{vmatrix}$$

$$= i \begin{vmatrix} 1 & 2 \\ 1 & 4 \end{vmatrix} - j \begin{vmatrix} -1 & 2 \\ 2 & 4 \end{vmatrix} + k \begin{vmatrix} -1 & 1 \\ 2 & 1 \end{vmatrix}$$

$$= 2\vec{i} - (-8)\vec{j} + (-3)\vec{k}$$

$$= \langle 2, 8, -3 \rangle$$

$$\vec{a} \cdot (\vec{b} \times \vec{c}) = \langle 1, 2, 3 \rangle \cdot \langle 2, 8, -3 \rangle$$

$$= 2 + 16 - 9 = 9$$

$$Volume = |\vec{a} \cdot (\vec{b} \times \vec{c})|$$

$$= |9| = 9$$

VECTORS: Ex. 7

Given: Two Vectors $\vec{p} = \langle 4, 8, 12 \rangle$ and $\vec{q} = \langle 10, 1, -3 \rangle$

Find: The vector projection of $\vec{p}$ onto $\vec{q}$

Solution:

Equation	$proj_q\,\vec{p} = \left(\dfrac{\vec{p} \cdot \vec{q}}{|\vec{q}|}\right)\dfrac{\vec{q}}{|\vec{q}|} = \left(\dfrac{\vec{p} \cdot \vec{q}}{|\vec{q}|^2}\right)\vec{q}$
Calculations	$\vec{p} \cdot \vec{q} = \langle 4, 8, 12 \rangle \cdot \langle 10, 1, -3 \rangle$ $\qquad = 40 + 8 - 36 = 12$
	$\|\vec{q}\| = \sqrt{10^2 + 1^2 + (-3)^2}$ $\|\vec{q}\| = \sqrt{110}$
Vector projection of $\vec{p}$ onto $\vec{q}$	$proj_q\,\vec{p} = \left(\dfrac{\vec{p} \cdot \vec{q}}{|\vec{q}|^2}\right)\vec{q}$ $= \left(\dfrac{12}{110}\right)\langle 10, 1, -3 \rangle$ $= \left(\dfrac{6}{55}\right)\langle 10, 1, -3 \rangle$ $= \langle \dfrac{60}{55}, \dfrac{6}{55}, \dfrac{-18}{55} \rangle$

Examples: Complex Numbers

COMPLEX NUMBERS: Ex. 1

Given: $y = 2x - 5$ (Rectangular Form)

Convert it from Rectangular Form to Polar Form.

Solution:	
Substitute: $x = r \cos \theta$ $y = r \sin \theta$	$y \;=\; 2x - 5$ $(r \sin \theta) \;=\; 2\,(r \cos \theta) - 5$
Solve for r	$r \sin \theta - 2r \cos \theta \;=\; -5$ $r\,(\sin \theta - 2 \cos \theta) \;=\; -5$ $r \;=\; \dfrac{-5}{(\sin \theta - 2 \cos \theta)}$
Note:	$r \;=\; f(\theta)$ That's the goal!

COMPLEX NUMBERS: Ex. 2	
Given: $r = 3\cos\theta$ (Polar Form) Convert it from Polar Form to Rectangular Form.	
Solution:	
Look for: $x = r\cos\theta$ $y = r\sin\theta$ $r^2 = x^2 + y^2$	$r = 3\cos\theta$ (Given) Can't find what I'm looking for!
Multiply both sides by r	$r = 3\cos\theta$ $r^2 = 3\,r\cos\theta$
Substitute	$x^2 + y^2 = 3x$
Rectangular Form	$x^2 + y^2 - 3x = 0$
Rectangular Form $y = f(x)$	$y^2 = 3x - x^2$ $y = \pm\sqrt{3x - x^2}$

Examples: Polar Curves

POLAR CURVES: Ex. 1

Given: $r = 5 + 4\cos\theta$

Graph the Polar Curve.

Solution:

Note:	Lined up along the x axis. Bean shape.

Get some (r, θ) data points Use $r = 5 + 4\cos\theta$	r	θ (Degrees)
	9	0
	5	90
	1	180
	5	270
	9	360

Graph	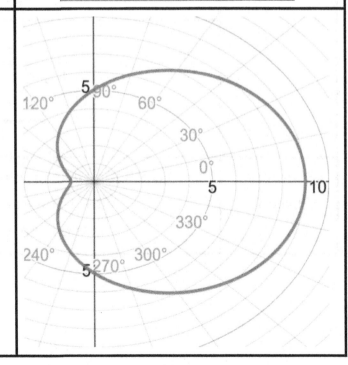

Examples: Polynomial Division

POLYNOMIAL DIVISION: General Information		
$\dfrac{Dividend}{Divisor} = Quotient$ ➔ $Divisor \overline{\big)\, Dividend}$ with $Quotient$ above		
There are two types of Polynomial Division. • Long Division • Synthetic Division:		
Long Division	Long division of polynomials is similar to regular long division of numbers.	
	$\dfrac{25}{4} = 6\dfrac{1}{4}$	$4\overline{)\,2\;5}$ quotient 6, $-\;2\;4$, 1 Remainder
Synthetic Division	Synthetic Division can only be used With 1st degree divisors in the form $(x - c)$	

POLYNOMIAL DIVISION: Ex. 1.

Use LONG DIVISION to evaluate: $\dfrac{x^3}{x^2 + 1}$

Solution:

Note: Every term must be represented so it lines up properly.

$$
\begin{array}{r}
x \\
x^2 + 0x + 1 \overline{\smash{\big)}\; x^3 + 0x^2 + 0x\ + 0} \\
-\ (x^3 + 0x^2 + 1x) \\
\hline
-x
\end{array}
$$

Remainder

$$
\dfrac{x^3}{x^2 + 1} \;=\; x - \dfrac{x}{x^2 + 1}
$$

POLYNOMIAL DIVISION: Ex. 2.

Use SYNTHETIC DIVISION to evaluate: $\dfrac{2x^3 + 5x}{x + 1}$

Solution:

Notes:

- Put the zero of the divisor in the box!
- Put coefficients of dividend in the top row.
- All terms must be represented. Use zeros to fill.

$$
\begin{array}{c|cccc}
 & x^3 & x^2 & x^1 & x^0 \\
\hline
-1 & 2 & 0 & 5 & 0 \\
 & \downarrow & -2 & 2 & -7 \\
\hline
 & 2 & -2 & 7 & \boxed{-7} \\
\end{array}
$$

Remainder

More Notes:

- The numbers in the bottom row represent the coefficients of the reduced polynomial.
- The 3rd degree polynomial was reduced to 2nd degree.

$$\frac{2x^3 + 5x}{x + 1} = 2x^2 - 2x + 7 - \frac{7}{x + 1}$$

POLYNOMIAL DIVISION: Ex. 3.

Use LONG DIVISION to evaluate: $\dfrac{x^2 + x^3 - 2x - 5}{x - 3}$

Solution:

$$
\begin{array}{r}
x^2 \quad + 4x \quad + 10 \\
x - 3 \overline{\smash{)}\; x^3 \quad + x^2 \quad - 2x \quad - 5} \\
-(x^3 \; - 3x^2) \\
\overline{\qquad\qquad 4x^2 \; - 2x} \\
-(4x^2 \; - 12x) \\
\overline{\qquad\qquad\qquad 10x \; - 5} \\
-(10x \; - 30) \\
\overline{\qquad\qquad\qquad\qquad 25 \quad \text{Remainder}}
\end{array}
$$

$$\dfrac{x^2 + x^3 - 2x - 5}{x - 3} = x^2 + 4x + 10 + \dfrac{25}{x - 3}$$

POLYNOMIAL DIVISION: Ex. 4.

Use SYNTHETIC DIVISION to evaluate: $\dfrac{x^3 + x^2 - 2x - 5}{x - 3}$

Solution:

Notes:
- Put the zero of the divisor in the box!
- Put coefficients of dividend in the top row.

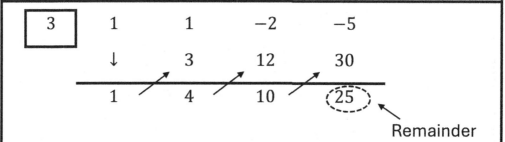

Remainder

More Notes:

- The numbers in the bottom row represent
- The coefficients of the reduced polynomial.

$$\frac{x^2 + x^3 - 2x - 5}{x - 3} = x^2 + 4x + 10 + \frac{25}{x - 3}$$

POLYNOMIAL DIVISION: Ex. 5.

Use LONG DIVISION to evaluate: $\dfrac{2x^4 - x^3 + 4x^2 - 14x + 6}{2x - 1}$

Solution:

$$
\begin{array}{r}
x^3 \qquad\qquad\quad + 2x \qquad - 6 \\
\hline
2x-1 \,\big)\ 2x^4 \quad - x^3 \quad + 4x^2 \quad - 14x \qquad + 6 \\
-(2x^4\ -\ x^3) \\
\hline
0 \quad + 4x^2 \quad - 14x \\
-(4x^2\ -\ 2x) \\
\hline
-12x \quad + 6 \\
-(-12x\ +\ 6) \\
\hline
0
\end{array}
$$

Remainder

$$\frac{2x^4 - x^3 + 4x^2 - 14x + 6}{2x - 1} \;=\; x^3 + 2x - 6$$

POLYNOMIAL DIVISION: Ex. 6.

Use SYNTHETIC DIVISION to evaluate: $\dfrac{2x^4 - x^3 + 4x^2 - 14x + 6}{2x - 1}$

Solution:

$$\frac{2x^4 - x^3 + 4x^2 - 14x + 6}{2\left(x - \frac{1}{2}\right)} = \frac{x^4 - \frac{1}{2}x^3 + 2x^2 - 7x + 3}{\left(x - \frac{1}{2}\right)}$$

Notes:

- Recall: Divisor must be in the form: $(x - c)$
- Put the zero of the divisor in the box!
- Put coefficients of dividend in the top row.

$$
\begin{array}{c|ccccc}
\frac{1}{2} & 1 & -\frac{1}{2} & 2 & -7 & 3 \\
 & \downarrow & \frac{1}{2} & 0 & 1 & -3 \\
\hline
 & 1 & 0 & 2 & -6 & 0 \\
\end{array}
$$

Remainder

More Notes:

- The numbers in the bottom row represent
- The coefficients of the reduced polynomial.

$$\frac{2x^4 - x^3 + 4x^2 - 14x + 6}{2x - 1} = x^3 + 2x - 6$$

POLYNOMIAL DIVISION: Ex. 7.

Solve with LONG DIVISION:
$$\frac{x^2 + 2x^2y - 2xy + 2xy^2 - 3y^2}{x + y}$$

Solution:

$$
\begin{array}{l}
\ x \qquad\qquad\ \ + 2xy \qquad\qquad\ - 3y \\
x+y\,\overline{\big)\ x^2 \quad + 2x^2y \ - 2xy \quad + 2xy^2 \quad - 3y^2} \\
\ -(x^2 \qquad\qquad + xy\,) \\
\ \overline{\qquad\quad 2x^2y \ - 3xy \quad + 2xy^2 \quad - 3y^2} \\
\ \qquad -(2x^2y \qquad\quad + 2xy^2\,) \\
\ \qquad\overline{\qquad\quad\ -3xy \qquad\qquad - 3y^2} \\
\ \qquad\qquad -(-3xy \qquad\quad - 3y^2\,) \\
\ \qquad\qquad\overline{\qquad\qquad\qquad\qquad\ 0}
\end{array}
$$

$$\frac{x^2 + 2x^2y - 2xy + 2xy^2 - 3y^2}{x + y} \;=\; x + 2xy - 3y$$

POLYNOMIAL DIVISION: Ex. 7.

Use SYNTHETIC DIVISION: $\dfrac{x^2 + 2x^2y - 2xy + 2xy^2 - 3y^2}{x + y}$

Solution:

$$= \dfrac{x^2\,(1 + 2y)\ +\ x\,(-2y + 2y^2)\ -\ (3y^2)}{x + y}$$

Notes:
- Treat "y" as a constant.
- Put the zero of the divisor in the box!
- Put coefficients of dividend in the top row.

$$\boxed{-y} \qquad (1 + 2y) \qquad (-2y + 2y^2) \qquad (-3y^2)$$

$$\qquad\qquad\qquad \downarrow \qquad\qquad\quad -y - 2y^2 \qquad\qquad 3y^2$$

$$\rule{9cm}{0.5pt}$$

$$\qquad\qquad 1 + 2y \qquad\qquad\qquad -3y \qquad\qquad\qquad 0$$

$$(1 + 2y)x\ -\ 3y\ =\ x + 2xy - 3y$$

More Notes:

- The numbers in the bottom row represent
- The coefficients of the reduced polynomial.

$$\dfrac{x^2 + 2x^2y - 2xy + 2xy^2 - 3y^2}{x + y}\ =\ x + 2xy - 3y$$

PART 3 – REINFORCE

Keeping important math equations in sight helps students learn them. Teachers and parents often place maps on walls to help students become more familiar with geography. In a similar way, teachers, parents and children may wear and display math equations to keep them in sight. Math is a beautiful thing!

Most of the math summaries are available on t-shirts, and tote bags. To find the t-shirts, just search for "Kathryn Paulk t-shirts" and a similar search for tote bags.

A few images of the t-shirts and are included in the following pages. The t-shirts are available in many colors for sizes for men, women, and children.

Math T-Shirts

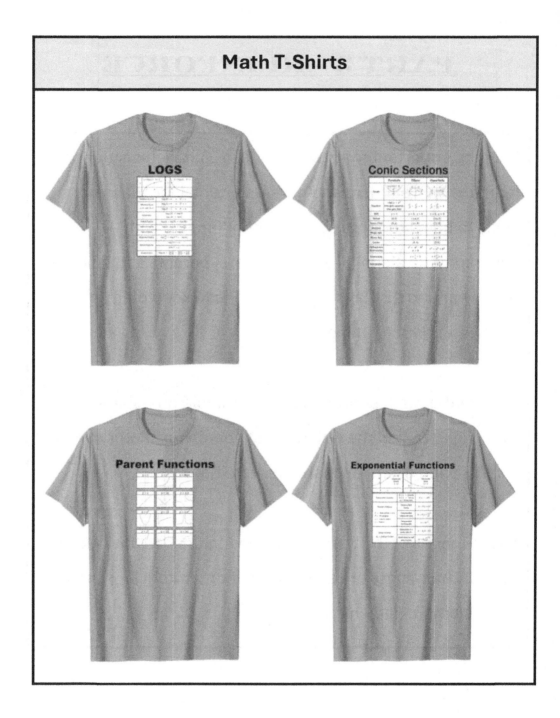

Other Amazon Books
By Kathryn Paulk

Other Books by Kathryn Paulk

- One-Page Summaries for Algebra, Geometry, and Pre-Calculus

- Graphing Functions Using Transformations for Algebra & Pre-Calc.

- Complex Numbers and Polar Curves For Pre-Calc and Trig: With Problems and Detailed Solutions

- Calculus 1 Review in Bite-Size Pieces

- Calculus 2 Review in Bite-Size Pieces

- Calculus 3 Review in Bite-Size Pieces

- Differential Equations With Applications: Class Notes With Examples

- Discrete and Continuous Probability Distributions: A Creative Comparison (V2)

- Teach Your Child to SWIM

BIG MATH For Little Kids

Workbooks for young children
& Solution manuals for parents

- Introduction to Numbers

- Introduction to Fractions
  by Sharing Things

- Introduction to Counting and Fractions
  by Cooking Breakfast

- Learn About Fractions   *****
  by Baking Cookies

- Adding Big Numbers, Guessing Numbers
  and Secret Codes

- Learn to Graph by Riding Bikes
  on Graph Paper

Made in the USA
Las Vegas, NV
04 February 2024

85307878R00083